"John Savino and Marie Jones have desc [...] volcanic explosions in human history. They go on to show how other potentially dangerous volcanoes such as Yellowstone and Long Valley need to be monitored, and preparations made in case the worst should happen. An extraordinarily detailed and compelling read."

—Simon Warwick-Smith, coauthor of *The Cycle of Cosmic Catastrophes*

"The Toba supervolcano eruption has been one of the most decisive events in the past 100,000 years, for the human race as well as for our green and dangerous planet. A serious and informative book on thesubject has been long overdue and necessary."

—George Weber, president ,The Andaman Association

"Using scientific analysis, Supervolcano exposes the massive catastrophes that are part of our geologic past. It's a lively detective story revealing how these events have profound repercussions on our species. With their book, Savino and Jones enliven the discussion on how we came to be who we are and what lies in our future."

—Joseph Christy-Vitale, author of *Watermark: The Disaster That Changed the World and Humanity 12,000 Years Ago*

"Who we are and where we come from are questions that never seem to go away. Painting a picture of what happened so long ago to make us who we are has always been a giant jigsaw puzzle for the historical disciplines. In Supervolcano, geophysicist Dr. John Savino and Marie Jones piece together that puzzle and paint that picture. From geology to genetics, and a touch of comparative mythology, they tell the story of mankind's earliest history in a savvy tale of discovery and science. Ultimately, they link our history to the catastrophic history of our planet, and specifically to a global cataclysm 75,000 years ago, which only a few thousand people survived. History and catastrophe, at times, are one and the same. For anyone interested in the origins, or the possible future of mankind, Supervolcano is a must read!"

—Edward F. Malkowski, author of *Before The Pharaohs*, and *The Spiritual Technology of Ancient Egypt*

SUPERVOLCANO

THE CATASTROPHIC EVENT THAT CHANGED THE COURSE OF HUMAN HISTORY (COULD YELLOWSTONE BE NEXT?)

DR. JOHN SAVINO, PH.D.

AND MARIE D. JONES

New Page Books
A Division of The Career Press, Inc.
Franklin Lakes, N.J.

SUPERVOLCANO
EDITED AND TYPESET BY GINA TALUCCI
Cover design by Howard Grossman/12e Design
Printed in the U.S.A. by Book-mart Press

To order this title, please call toll-free 1-800-CAREER-1 (NJ and Canada: 201-848-0310) to order using VISA or MasterCard, or for further information on books from Career Press.

The Career Press, Inc., 3 Tice Road, PO Box 687,
Franklin Lakes, NJ 07417
www.careerpress.com
www.newpagebooks.com

Library of Congress Cataloging-in-Publication Data
Savino, John.
    Supervolcano : the catastrophic event that changed the course of human history / By Dr. John Savino, PhD. and Marie D. Jones.
        p. cm.
    ISBN-13: 978-1-56414-953-4
    ISBN-10: 1-56-414-953-6
    1. Volcanoes. 2. Volcanoes—History. 3. Emergency management. I. Jones, Marie D. II. Title.

QE522.S28 2007
551.21--dc22

                                                                    2007017190

Dedication

For Max, the boy who loves volcanoes.
—Marie

Thank you, Susan, for your patience and support.
—John

Acknowledgments

We would like to thank Lisa Hagan, agent extraordinaire, for her belief in this project. We would like to thank Michael Pye and the entire staff at New Page Books for their support and enthusiasm. You guys rock! We would also like to thank all of the scientists, photographers, and researchers whose work is featured in this book, including but not limited to Stanley Ambrose, Stephen Self, Michael Rampino, George Weber of the Andaman Assocation, Bill McGuire, and all the great men and women at USGS working to understand the nature of the mighty supervolcano.

John would also like to thank…

his wife, Susan, for her understanding, patience, and support whenever he went into seclusion to work on this book.

Thanks go to Rick Otto, superintendent at the Ashfall Fossil Beds State Historical Park for his help in getting a picture of ancient mammals at the watering hole in Nebraska 10 million years ago.

Marie would also like to thank…

Lisa Collazo, of *Writewhatyouknow.com*, the best writing coach out there. And thanks so much to the following friends and family for their love, support, ideas, and inspiration: Milly Savino, Ron Jones, Angela Gonzalez and family, John and Winnie Savino, Andrea Glass, Helen Cooper (onward and upward!), Sharon Schulz Elsing, Michael DiBernardo, Marit Flowers, Laura Jones, and all my wonderful PDSD colleagues. And of course, Max, for making it all worthwhile.

Contents

INTRODUCTION

The Dragon Sleeps

Why would a volcanic eruption that occurred approximately 74,000 years ago matter so much to those of us alive today? What possible connection could we have with those who walked the Earth so many thousands of years ago? There have been volcanic, even super-volcanic eruptions throughout the history of our planet, yet what happened in the center of the northernmost part of the Indonesian island of Sumatra sometime between 70,000 and 75,000 years ago, at a volcano called Toba, literally changed the course of human evolution in a way that no other singular event has in recent geological history. At least not since human footprints have been found upon the Earth.

Toba is not your garden-variety volcano. It is a supervolcano. The volume of material erupted by Toba is estimated to be about 2,800 times greater than the material spewed out during the relatively puny 1980 Mount St. Helens eruption. Even the largest eruption in recorded history, Tambora in Indonesia in 1815, which altered the climate of the entire globe and killed more than 90,000 people, pales in comparison to the mighty supervolcano. Toba, by contrast, ejected about 300 times more volcanic ash.

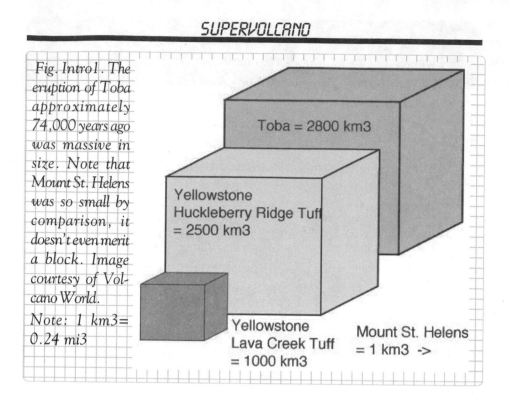

Fig. Intro1. The eruption of Toba approximately 74,000 years ago was massive in size. Note that Mount St. Helens was so small by comparison, it doesn't even merit a block. Image courtesy of Volcano World.

Note: 1 km3 = 0.24 mi3

Toba = 2800 km3

Yellowstone Huckleberry Ridge Tuff = 2500 km3

Yellowstone Lava Creek Tuff = 1000 km3

Mount St. Helens = 1 km3 ->

But Toba is more than just a massive and amazing force of nature; when it erupted, it created a volcanic winter that plunged temperatures around the globe, causing environmental chaos. Toba changed us as human beings. It changed our species, our way of life, and the way we looked at the natural world around us, including the gods and goddesses we worshipped.

The story of Toba is one of change. When an event occurs that wipes out most of your own species, not to mention most of the plant and animal life you feed upon, you change. You change physically, as did the few lucky survivors who went on to become the genetic grandparents of all humanity. You change emotionally, as the 9/11 terrorist attacks changed all of us on some level. What happened to those living at the time of Toba was far more terrifying, for it took away much of what sustained life and erased the signature of every existing branch of humanity, except for one. Imagine losing everyone and everything you love, and everything you need for survival. It would change you, too.

But we believe you also change spiritually when a cataclysmic event of such magnitude rips your world, and the world itself, apart.

We wanted to write this book for many reasons:

✳ We, similar to millions of other people, find supervolcanoes simply fascinating.

✳ Toba was the biggest supereruption in the last 2 million years.

✳ Toba wiped out most living things on the planet.

✳ Toba created a rare population bottleneck that shifted the genetic footprints of the human species.

✳ Toba caused severe radiation poisoning to exposed humans via destruction of a portion of the ozone layer that shields the Earth from the sun's dangerous ultraviolet radiation.

✳ Toba created an environmental situation that led to global climate change—something we now struggle with today.

✳ Toba, or another mighty supervolcano such as Yellowstone in Wyoming or Long Valley in California, could erupt again. In both cases, there are intriguing signs of unrest.

✳ Natural disasters both thrill and repel us as a species; we want to look away, but we cannot. We are frightened, yet drawn like the moth to the flame.

But we really chose to tell this story because it has not been told before, and because it speaks of our innate human desire to understand where we came from, how we got here, and most importantly, where we are headed in the future. Could telling the tale of Toba prevent such disaster on an epic scale from happening again? Maybe…maybe not. Could documenting our genetic commonality stop wars and racism? Maybe…maybe not. Could scaring people into a state of preparedness keep them from suffering the same fate as Toba's many victims? Maybe…maybe not.

We chose to tell the story anyway.

As two people who deeply respect the lessons of the historical past, the knowledge of the scientific world, and the power of the written word to tell a story of mythological proportions that resonates on so many levels even today, with headlines of global climate change, deadly tsunamis, and new volcanoes found under the sea…we could not NOT write this story.

Ultimately, Toba is just a physical mark upon the Earth where right now a gorgeous, deceptively peaceful lake sits; a location on a map with little meaning for those not already aware of its critical impact on humanity. Standing upon its banks, you would think of how lovely it looked, never knowing you were looking into the maw of a dragon, whose underbelly still churns with magma and hot gasses, although much less violently than it did 70,000-odd years in the past. A dragon whose fiery breath has, in some way, touched every human on Earth, albeit at a cellular, genetic level.

But for those with the wisdom and the insight to look beyond the surface of the dragon, Toba is a deeper, more constant reminder of how fragile our existence is, and how easily it can be changed into something else for those lucky enough to survive.

Fig. Intro2. Lake Toba today, beautiful and serene. Image courtesy of Alan Ingram. Image courtesy of Alan Ingram (www.caingram.info)

That *we* survived. Toba is *our* story, and yours, too; for we are all made of the fiery breath of the same dragon. On the morning of December 26, 2004, the second-largest earthquake in the past 100 years occurred off the coast of northern Sumatra, creating a massive tsunami that devastated the region. This mega-thrust event ruptured approximately 745 miles of the Indonesian subduction zone. The epicenter of the earthquake was about 250 miles west of Lake Toba. Was this the stirrings of the dragon? We will explore the relationship of the great December 26th earthquake and the volcanic zone in that region in a subsequent chapter.

But for now the question remains: Which dragon will next stir and breathe its fiery breath upon the world in the form of a supervolcanic eruption? Only time will tell.

PART I

Supervolcano

CHAPTER 1

Mountains of Fire

From the moment of its formation, the Earth has been restless. Natural disasters of cataclysmic proportions not only helped shape and mold the planet in its initial stages, but also continue to transform the face of the world we now live upon and call home.

Earthquakes, tsunamis, meteor impacts, and floods have all contributed to the ever-changing landscape. Even today we struggle to understand, and survive, massive movements of earth and water that often come without warning. But none have had a greater effect than the mighty volcano. And of the hundreds of volcanoes that dot the Earth, many clustered near boundaries of the Earth's major tectonic plates, none are more powerful and transformative than the supervolcano.

Television and the Internet have brought us to the edge, where we can literally peer down into the belly of a volcano. We can sit back in our recliners and watch molten rock spew over the landscape (lava), and scenes of a dense, hot, chaotic avalanche of rock fragments, gas, and volcanic ash race away from an explosive eruption column (pyroclastic flow) at speeds greater than 60 mph, devastating everything in their path. We were all awe-struck when live images of eruptions of Mount St. Helens in Washington State and Pinatubo in the Philippines kept us glued to our television sets. But these and other highly publicized eruptions are no match for the one force of nature that has the ability to truly change the face of the entire planet, and the living things lucky enough to survive. That force of nature is the supervolcano!

WHAT IS A SUPERVOLCANO?

To qualify as a supervolcano, the volcano itself must be large enough to result in a supereruption of ejected material more massive than any volcano in the historical record. This colossal eruption doesn't just consist of huge incandescent hurricanes of hot gasses, ash, and rocks, which can cover thousands of square miles and obliterate all forms of life in its path, but supereruptions have the potential to do far more damage. In other words, they are judged by the global extent of their catastrophic effects on the environment and, as we will see, civilization.

The London Working Group of the Geological Society, in their chilling 2005 report, "Super-eruptions: global effects and future threats," states that "many large volcanoes on Earth are capable of explosive eruptions much bigger than any experienced by humanity over historic time." The Working Group, comprised of six renowned earth scientists, goes on to claim, "Super-eruptions are different from other hazards such as earthquakes, tsunamis, storms or floods in that—like the impact of a large asteroid or comet—their environmental effects threaten global civilization."

Another sobering fact is that there is no known mechanism for averting a devastating supereruption. This is in contrast to the possibility of sufficiently perturbing the orbital parameters of an incoming asteroid or other near-Earth object to avoid a collision with the Earth, a possibility being actively pursued by the European Space Agency under Project Don Quijote. The Project adopted the Spanish spelling of Don Quixote, the protagonist in Cervantes' novel, who has chivalrous ideas that tend toward the fanciful and impractical. The National Aeronautics and Space Administration is also pursuing a mission, known as the Deep Impact mission, a six-year project that culminated on July 3, 2005, with the controlled collision of a small "impactor" satellite with the comet Tempel 1.

The mechanism, tectonic settings, frequency of occurrence, and environmental and climatological effects of supervolcanoes will be discussed later. The last known supervolcano erupted at Lake Taupo on the North Island of New Zealand about 26,500 years ago. Any prehistoric people who might have been close enough to see the first few seconds of the eruption would have ceased to exist shortly thereafter. The complete lack of any kind of direct observation forces us to turn to information about the regional and global environmental impacts of historic volcanic eruptions. These can hopefully provide insight and guidance for estimating the aftermath of a supervolcano on the scale of the largest known volcanic event in possibly the past 27 million years—Toba.

PARÍCUTIN

Before we get into the nuts and bolts of volcanoes, let's take a look at an event of special significance in the world of volcanology that happened in 1943 in a cornfield in Mexico. The significance of this event is that it provided a reality check for volcanologists in their understanding of the formation and cessation of volcanic activity. They were able to witness the event firsthand, and make detailed observations of the complete life-cycle of a volcano.

On Saturday afternoon, February 20, 1943, not far from the Mexican village of Parícutin, Dionisio Pulido, a Tarascan Indian, was burning shrubbery in his cornfield in preparation for spring sowing. His wife and son were nearby shepherding their young lambs when, suddenly, there was a roar of thunder and the ground in front of them tore open to form a fissure about 150 feet long. Dionisio later recalled that he and his family heard loud and continuous hissing or whistling sounds and saw smoke rising from one end of the fissure, which smelled like "rotten eggs." The rotten egg smell is the characteristic odor of hydrogen sulfide commonly associated with volcanic eruptions and hot springs worldwide. Dionisio and his family had just witnessed the birth of a volcano.

By the next morning, the young volcano had taken on a conical shape, and had grown to a height of 30 feet. During the day the volcano grew another 120 feet, and had taken on the classic form of a scoria cone and that night incandescent bombs blew out from the top more than 1,000 feet up into the darkness. Unfortunately for Dionisio and his family, a slag-like mass of lava spilled out of the young volcano and encompassed their cornfield. The volcano was later named after the nearby village, Parícutin, even though lava flows from the growing volcano would end up destroying the village just a few months later.

News of the birth of Parícutin spread rapidly, and within a few days geologists and volcanologists from many parts of the world came to study this extraordinarily unique volcanic event. The volcano remained active for nine years, from 1943 to 1952, and marked the first time scientists were able to observe the life cycle of a volcano from birth to extinction.

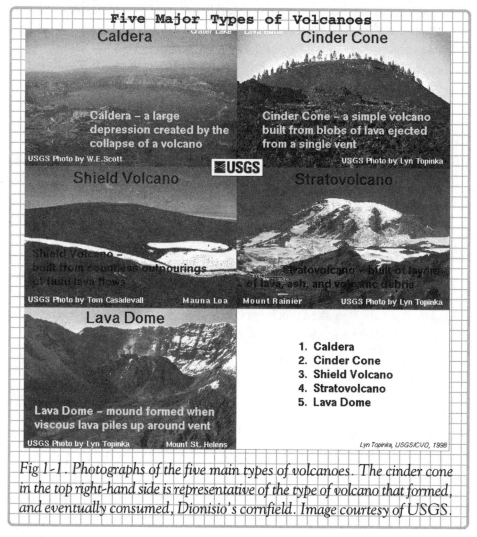

Five Major Types of Volcanoes

Caldera
Caldera – a large depression created by the collapse of a volcano
USGS Photo by W.E.Scott

Cinder Cone
Cinder Cone – a simple volcano built from blobs of lava ejected from a single vent
USGS Photo by Lyn Topinka

USGS

Shield Volcano
Shield Volcano – built from countless outpourings of fluid lava flows
USGS Photo by Tom Casadevall — Mauna Loa

Stratovolcano
Stratovolcano – built of layers of lava, ash, and volcanic debris
Mount Rainier — USGS Photo by Lyn Topinka

Lava Dome
Lava Dome – mound formed when viscous lava piles up around vent
USGS Photo by Lyn Topinka — Mount St. Helens

1. Caldera
2. Cinder Cone
3. Shield Volcano
4. Stratovolcano
5. Lava Dome

Lyn Topinka, USGS/CVO, 1998

Fig 1-1. Photographs of the five main types of volcanoes. The cinder cone in the top right-hand side is representative of the type of volcano that formed, and eventually consumed, Dionisio's cornfield. Image courtesy of USGS.

The Parícutin eruption was unusually long for what is known as a Strombolian eruption, with several eruptive phases occurring over its nine-year life span. For the first two years, pyroclastic activity was the dominant eruptive mode. This activity then waned and was replaced by an outpouring of lava from the base of the cone for the remaining seven years. The only deaths associated with the volcano were three people who were struck by lightning generated during the early pyroclastic eruptions. The final height of the scoria cone was estimated to be about 1,390 feet.

The preceding account of the life cycle of the volcano, Parícutin, introduced several terms in volcanology. For instance, we saw that Parícutin was a scoria or cinder cone built from particles and blobs of congealed lava ejected from a single vent. Scoria or cinder cones are one of several main types of volcanoes. These include, in order of increasing eruptive intensity or violence: shield volcanoes, scoria cones, stratovolcanoes, lava dome, and calderas.

VOLCANIC EXPLOSIVITY INDEX

Before proceeding with descriptions of these different types of volcanoes, we introduce a relatively simple, semi-quantitative scheme developed in 1982 by Chris Newhall of the U.S. Geological Survey (USGS) and Steve Self, then at the University of Hawaii (presently at the Open University [UK]), for estimating the magnitude of historic eruptions. The scheme is referred to as the Volcanic Explosivity Index (VEI). Historical eruptions can be assigned a VEI number on a scale of 0 to 8 based on one or more of the following criteria:

* Volume of ejected material.
* Eruptive column height.
* Subjective descriptions of the eruption (gentle, effusive, explosive, cataclysmic, and so on).
* Plume head height.

The volume of ejected material and the plume height are considered to be the two most reliable criteria in assigning a VEI number. The VEI scale is analogous to the Richter magnitude scale for estimating the size of earthquakes, in that both are logarithmic scales wherein an increase of 1 unit on either scale implies an order of magnitude (that is, a factor of 10) increase in the intensity (amplitude of a seismic wave, volume of volcanic ejecta) of the corresponding seismic or volcanic event. Based on a compilation of VEI numbers for Holocene (last 10,000 years) eruptions, Parícutin was assigned a VEI of 4.

Scientists from Cambridge University performed an investigation of 47 of the largest explosive eruptions on Earth, all with VEIs of 8 or greater, ranging in age from 26,500 years old to 454 million years old. As part of their investigation, the scientists devised a scale for comparing sizes of supereruptions that mitigates one of the shortcomings of the VEI scale. The problem with the VEI approach is that it does not take into account the different densities of material ejected from a volcano. This can result in a high VEI number for an eruption that produces a large volume of fluffy ash compared to an eruption that produces a smaller volume of dense volcanic rock. Based on the VEI scheme, the first fluffy event might score a higher VEI number than the event that consists of denser volcanic rock, hardly a fair outcome. The results of the Cambridge study, especially as they pertain to the temporal behavior of supereruptions, will be discussed in greater detail in Chapter 4 when we focus on the makeup of supervolcanoes.

THE MAKING OF A TYPICAL VOLCANO

Now, let's take a look at a likely scenario for what is going on deep inside the Earth that led up to the volcanic eruption and formation of a cinder cone in Dionisio's cornfield. Magma, or molten rock, forms deep in the earth and is propelled upward by buoyancy, because the magma is less dense than the surrounding deep source, or host rock. The buoyant force operative here is based on the principle discovered by Archimedes (287 to 212 B.C.) while, according to some, getting into his bathtub and noticing how the water level rose. The story goes that he was so excited with his discovery that he jumped up and ran into the street shouting, "Eureka!" The only problem was that, in his state of euphoria, he had forgotten to get dressed and was in his birthday suit!

Meanwhile, back underground, as the magma rises it moves into colder and harder rock where it slows down and collects in relatively shallow chambers. At the shallower depths, gases, some of them pretty nasty, which were previously dissolved in the magma, begin to come out of solution. The combination of buoyancy and degassing increases the pressure on the confining rock until, similar to a champagne bottle popping its cork, the magma erupts through a vent or, as in the Dionisio cornfield, a fissure, sending igneous rock, ash, and small pebbles known as lapilli into the air. Smoke containing steam, carbon dioxide, ash, and highly toxic sulfur and halogen (chlorine and fluorine) gases, including the hydrogen sulfide that Dionisio and his wife identified with "rotten eggs," is released into the atmosphere during an eruption.

Meanwhile, liquid volcanic rock called lava (it's called magma while in the belly of a volcano, and lava once it is released on the surface), at a temperature between 1,500 and 2,200°F, flows out of vents in the volcano and spills down the mountainside, melting everything in its unfortunate path. The eruptive type can vary from highly explosive (Mount Pinatubo in the Philippines on June 15, 1991), to one of the frequent effusive outpourings of lava (Kilauea on the Big Island of Hawaii). Repeated eruptions during long periods of time form the mountains or mountain-like features that we, and Dionisio and his family, associate with the notion of "volcano."

The eruptive behavior of a volcano is strongly dependent on the composition of the magma that pools in a chamber beneath the volcano. Magma is a complex high-temperature (between 1,500 and 2,200°F) silicate solution with varying amounts of volatiles (water and gases), all under high pressure. Important properties of the magma are its viscosity, or resistance to fluid flow, and the amount of dissolved gas in the melt. Basically, many years of investigations have shown that, in general, the higher the viscosity, the more explosive the resultant eruption. Compositionally, this is explained by the observation that increasing the silica content of magma increases the viscosity of the magma, which results in the increased explosiveness of an ensuing eruption. This is somewhat of an oversimplification, as there are obviously other physical and compositional properties of the magmatic mix that determine the eruption dynamics. These other properties will be addressed in Chapter 4, when the life cycle of a supervolcano is examined in more detail.

TYPES OF VOLCANOES

The following figure summarizes the types, principal characteristics, some examples of corresponding volcanoes, and resultant landforms of eruptions. The progression is from relatively nonexplosive, low silica, low viscosity effusive outpourings of lava (flood basalts), to the high silica, high viscosity, largest known eruptions to have ever occurred (caldera eruptions including supervolcanoes). In terms of the VEI scale, the violence or explosivity increases from VEI 0 for the activity associated with the shield volcanoes (these volcanoes are occasionally assigned a VEI rating of 1), to VEI 8 for the caldera eruptions.

Types of Volcanoes

Volcano Type	Characteristics	Examples	Simplified Diagram
Flood or Plateau Basalt	Very liquid lava; flows very widespread; emitted from fractures	Columbia River Plateau	1 mile:
Shield Volcano	Liquid lava emitted from a central vent; large; sometimes has a collapse caldera	Larch Mountain, Mount Sylvania, Highland Butte, Hawaiian volcanoes	
Cinder Cone	Explosive liquid lava; small; emitted from a central vent; if continued long enough, may build up a shield volcano	Mount Tabor, Mount Zion, Chamberlain Hill, Pilot Butte, Lava Butte, Craters of the Moon	
Composite or Stratovolcano	More viscous lavas, much explosive (pyroclastic) debris; large, emitted from a central vent	Mount Baker, Mount Rainier, Mount St. Helens, Mount Hood, Mount Shasta	
Volcanic Dome	Very viscous lava; relatively small; can be explosive; commonly occurs adjacent to craters of composite volcanoes	Novarupta, Mount St. Helens Lava Dome, Mount Lassen, Shastina, Mono Craters	
Caldera	Very large composite volcano collapsed after an explosive period; frequently associated with plug domes	Crater Lake, Newberry, Kilauea, Long Valley, Medicine Lake, Yellowstone	

Increasing Violence / Increasing Viscosity

USGS

Topinka, USGS/CVO, 1997, Modified from: Allen, 1975, Volcanoes of the Portland Area, Oregon, Ore–Bin, v.37, no.9

Fig. 1-2. Major types of volcanoes and some of their principal characteristics. Volcanic landforms shown in order of increasing viscosity and violence (see arrows on left-hand side of figure) from top to bottom.

FLOOD BASALTS

The first column in the figure refers to the volcano types, starting with the Flood or Plateau Basalt, and ending at the bottom of the figure with caldera. The Flood Basalts are considered one of two types of supervolcanoes because of the exceptionally large amount of the eruption material they produce. While supposedly primarily effusive in nature, the continental Flood Basalts can result in basaltic provinces covering several million square miles, and have volumes on the order of one-half million or more cubic miles.

Approximately 250 million years ago, during the golden era of biodiversity known geologically as the Permian era, one of the biggest volcanic events ever to have occurred on Earth covered extensive portions (several million square miles) of primeval Siberia. Some scientists believe that this event and the resulting basalt formation, known as the Siberian Traps, were the cause of the largest mass extinction ever, during which up to 95 percent of ocean life forms became extinct.

SHIELD VOLCANOES

The next major type of volcano is the shield volcano. This type is probably best exemplified by the Hawaiian chain of volcanoes, which are believed to have been produced as the Pacific tectonic plate passed over a hot spot. Supporting this view is the observation that in general, as you move along the island chain from the Big Island of Hawaii to Kauai, the volcanoes become progressively older. The progression of older ages continues to the Kure Atoll 1,500 miles northwest of the Big Island. Kure Atoll is a low coral island marking the location of an extinct volcano that was active about 30 million years ago. Approximately 400 miles northwest of Kure, the chain of submerged extinct volcanoes, known as the Emperor seamounts, suddenly trends for another 1,700 miles in a more northerly direction to the Aleutian trench. Whether this change is the result of a change in the direction of motion of the Pacific plate about 40 million years ago or a moving hotspot is a hotly debated issue. Stay tuned!

Mauna Loa is an impressively large mountain in all aspects. At least 60 miles long and 30 miles wide, it makes up at least half of the Big Island of Hawaii. Considering that the flanks of Mauna Loa sit on sea floor that is about 16,400 feet deep, the "height" of this volcano relative to the sea floor is about 30,080 feet, making it the largest active volcano in the world. In fact, using this last measure of height, it is one of the tallest mountains in the world (although many mountains, such as Mt. Everest in the Himalaya mountain range, sit higher relative to sea level).

The massive size of Mauna Loa has an effect on the sea floor beneath it that adds to the height of this monster volcano. All large land masses (such as mountains) push down upon the Earth's crust due to their enormous weight. So, directly beneath Mauna Loa, the sea floor on which it sits is depressed by an estimated additional 26,000 feet. Thus, if one wanted to estimate the thickness at the center of the lava pile that makes up Mauna Loa, one would need to add its above sea level height, its sea floor to sea level height, and the amount of the depression in the Pacific sea floor caused by the weight of Mauna Loa. These dimensions add up to a total of 56,080 feet.

The ongoing volcanic activity of the Big Island reflects the fact that the Island currently sits over the hotspot that is presumably beneath the very active Kilauea. Five major volcanoes comprise the Big Island: Kilauea, Mauna Loa, Mauna Kea, Hualalai, and Kohala. While Mauna Loa is the largest active volcano on Earth, Kilauea is one of the most productive. An interesting development is the formation of the youngest volcano in the chain, Loihi, about 15 miles southeast of Kilauea. Loihi is properly referred to as a seamount because it hasn't poked its head above the ocean surface, and is not expected to do so for tens of thousands of years.

CINDER CONE

The Parícutin cinder or scoria cone volcano discussed earlier is known as a Strombolian eruption. A Strombolian eruption is relatively nonviolent, and is named after the mild to moderate eruptions that occur at intervals ranging from minutes to hours on the island of Stromboli. The island is known as the "Lighthouse of the Mediterranean," because for at least the past 1,000 years, the somewhat regular occurrence of incandescent nighttime explosions have warned sailors of the proximity and potential hazards of shallow waters.

Between 1888 and 1890, Guiseppe Mercalli, the Italian seismologist known for his compilation of the 12-level Mercalli scale of earthquake damage, witnessed several eruptions on the island of Vulcano. He described the vulcanian eruptions as resembling "cannon firings at long intervals." Mercalli's description of the eruption style is now used all over the world to characterize the moderately explosive eruptions. The vulcanian type eruption has been found to characterize eruptions associated with more explosive scoria cones and mildly explosive stratovolcanoes.

STRATOVOLCANOES

More explosive stratovolcanoes are characterized by an eruptive style referred to as Plinian. This style of eruption is named after Pliny the Younger, a Roman statesman who wrote a vivid account of the highly explosive eruption in A.D. 79 of Mt. Vesuvius. This eruption is responsible for the deaths of thousands of people, and buried the Roman towns of Pompeii and Herculaneum under huge volumes of tephra, pyroclastic flows, and lahars. Pompeii remained buried for more than 1,700 years until it was accidently discovered during the excavation of a water line. The uncovering of Pompeii by archeologists has greatly increased our understanding of Plinian-style eruptions, as well as the lives of ordinary people during Roman times.

While the eruptions of stratovolcanoes usually capture the headlines instead of the shield volcanoes eruptions in Hawaii, it is interesting to compare the dimensions of these two types of volcanoes. Recall that the Hawaiian shield volcano Mauna Loa was identified as the largest volcano on Earth. This is shown somewhat dramatically in the following figure where it is compared with Mount Rainier, one of the larger Cascade Range composite volcanoes.

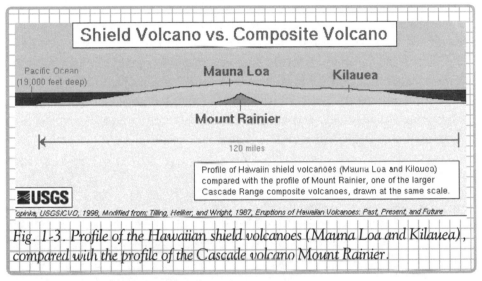

Fig. 1-3. Profile of the Hawaiian shield volcanoes (Mauna Loa and Kilauea), compared with the profile of the Cascade volcano Mount Rainier.

VOLCANIC DOME

The next type of volcano is the volcanic or lava dome, which forms when lava reaching the surface of the Earth is so viscous that it cannot flow readily, and, therefore, accumulates around the vent. One of the largest known volcanic domes is that constituting the upper part of Lassen Peak in northern California. The Lassen dome rises more than 2,000 feet, and has a diameter of approximately 2 miles.

CALDERA

The last volcano type is the caldera. When a volcanic eruption depletes the magma in a shallow-level magma chamber, the volcanic edifice can collapse under the force of gravity into the empty chamber. In this way, a caldera

(Spanish for kettle or cauldron) is formed typically as a steep, bowl-shaped depression. Calderas are highly variable in size, with diameters ranging from about 6 to 60 miles. Variations in types of calderas will be explored in greater detail in Chapters 4 and 5 with special attention to the type or class of calderas referred to as resurgent. It is among this class of calderas that we encounter the biggest supervolcanoes, in particular, Toba.

WHERE ARE VOLCANOES FOUND?

The source regions of volcanoes in general and supervolcanoes in particular fall into two categories: subduction zones and mantle hot spots. In general, supervolcanoes are more likely to be found associated with subduction zones—places around the Earth where the interaction of tectonic plates results in areas of mega-thrust (an oceanic plate dips below a continental plate). Many of the most widely known, and active, subduction zones on the planet occur along the Pacific Rim, or Ring of Fire.

Major volcanic areas in the Ring of Fire include:

* In South America, the Nazca plate is colliding with the South American plate. This has created the Andes, and volcanoes such as Cotopaxi and Azul.

* In Central America, the tiny Cocos plate is crashing into the North American plate, and is therefore responsible for the Mexican volcanoes of Popocatepetl and Parícutin (which, as we saw earlier, rose up from Dionisio's cornfield in 1943 and became an instant mountain).

* Between Northern California and British Columbia, the Pacific, Juan de Fuca, and Gorda plates have built the Cascades, and the infamous Mount Saint Helens, which erupted in 1980.

* Alaska's Aleutian Islands are growing as the Pacific plate hits the North American plate. The deep Aleutian Trench has been created at the subduction zone with a maximum depth of 25,200 feet.

* From Russia's Kamchatka Peninsula to Japan, the subduction of the Pacific plate under the Eurasian plate is responsible for the Japanese islands and volcanoes (such as Mt. Fuji).

✳ The final section of the Ring of Fire exists where the Indo-Australian plate subducts under the Pacific plate, and has created volcanoes in the New Guinea and Micronesian areas. Near New Zealand, the Pacific Plate slides under the Indo-Australian plate.

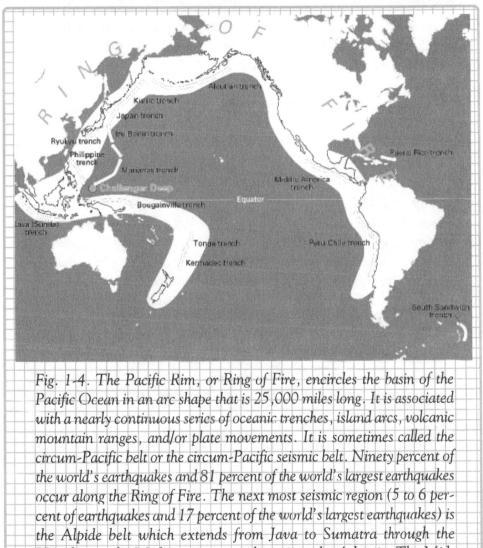

Fig. 1-4. The Pacific Rim, or Ring of Fire, encircles the basin of the Pacific Ocean in an arc shape that is 25,000 miles long. It is associated with a nearly continuous series of oceanic trenches, island arcs, volcanic mountain ranges, and/or plate movements. It is sometimes called the circum-Pacific belt or the circum-Pacific seismic belt. Ninety percent of the world's earthquakes and 81 percent of the world's largest earthquakes occur along the Ring of Fire. The next most seismic region (5 to 6 percent of earthquakes and 17 percent of the world's largest earthquakes) is the Alpide belt which extends from Java to Sumatra through the Himalayas, the Mediterranean, and out into the Atlantic. The Mid-Atlantic Ridge is the third most prominent earthquake belt. Courtesy of USGS.

THE SCIENCE BEHIND VOLCANOES

So far, we have only dabbled in the relationship between plate tectonics and volcanism. The Ring of Fire seen in the sidebar was known for years before the theory of plate tectonics was put forward in the mid-to-late 1960s and 1970s. Plate tectonics is a combination of two earlier ideas: continental drift and sea-floor spreading. Continental drift was formally proposed by Frank Bursley Taylor at a meeting of the Geological Society of America in 1908. He proposed that the continents moved on the Earth's surface and, somewhat amazingly, that a shallow region in the Atlantic marks the location where Africa and South America were once joined. What's amazing about Taylor's hypothesis is that the Mid-Atlantic Ridge, the shallow region in the Atlantic, wasn't discovered until the 1950s.

Taylor's ideas were picked up by Alfred Wegener and published in 1915 in his famous book, *The Origin of Continents and Oceans*. Wegener's hypothesis did not win wide support from the geology and geophysics community because he was unable to come up with a convincing mechanism for the drifting continents. In the late 1950s and early-to-mid 1960s, several scientists, including Robert Dietz, Harry Hess, and J. Tuzo Wilson, developed the hypothesis referred to as sea-floor spreading. This hypothesis stated that oceanic crust was created at mid-ocean ridges, and subsequently moved away from the ridge crests toward the deep trenches. Within a few years, the concepts of continental drift and sea-floor spreading were merged into the theory of plate tectonics.

According to this theory, the Earth's outermost layer, the lithosphere, is a relatively strong layer consisting of the crust and the rigid uppermost part of the Earth's mantle. There are two types of lithosphere: oceanic, which is about 30 to 62 miles thick, but thins considerably as the mid-ocean ridges are approached; and continental, which is typically about 90 miles thick. The lithosphere sits on top of a layer of the upper mantle referred to as the asthenosphere, which is a relatively hot, weak layer upon which the rigid and brittle lithospheric plates move about at rates measured in inches per year. The asthenosphere extends to depths of 62 to 125 miles, or more, below the Earth's surface.

The lithosphere is broken up into seven major and many minor tectonic plates, the largest being the Pacific plate. The lithospheric plates ride on the asthenosphere, and move relative to each other at fingernail growing speeds of generally 1/2 to 5 inches per year. The edges of these plates, where they run into, crash, and grind together are the sites of intense geologic activity, such as earthquakes, volcanoes, and mountain building. The driving force is presumably slowly moving convection currents within the mantle, which, in turn, are thought to be driven by heat from the Earth's core, similar to convection currents in a pot of water heated on a stove.

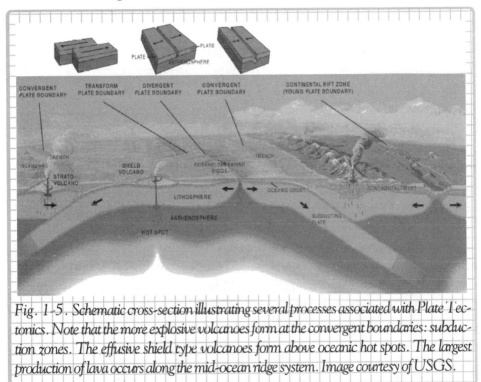

Fig. 1-5. Schematic cross-section illustrating several processes associated with Plate Tectonics. Note that the more explosive volcanoes form at the convergent boundaries: subduction zones. The effusive shield type volcanoes form above oceanic hot spots. The largest production of lava occurs along the mid-ocean ridge system. Image courtesy of USGS.

The overwhelming majority of subaerial volcanoes occur where tectonic plates converge—at subduction zones where two plates moving toward each other meet with one plate sliding underneath the other and moving down into the mantle. Zelle de Boer and Donald Sanders present an excellent description of the process of magma formation in a subduction zone in their book *Volcanoes in Human History: The Far-Reaching Effects of Major Eruptions*. These authors note that as the subducted plate descends into the asthenosphere, the high temperatures and pressures prevailing at depth force fluids out of the subducted rock. As the subducted plate approaches a depth of about 43 miles, it starts to release volatile gases. By the time the plate has reached a depth of 125 miles, all liquids and gases have been forced out of it by the high prevailing pressures. Thus, it is between depths of 60 and 90 miles that magma is generated. Blobs of magma rise slowly through the asthenosphere to the bottom of the lithosphere, where they coalesce into sheets of molten material hot enough to melt adjacent parts of the lithosphere. At this point the scenario beneath Dionisio's volcano, prior to the eruption, takes over and the material eventually forces its way through fractures to the Earth's surface.

Volcanoes forming along the subduction zones are typically explosive, and. as mentioned previously, this is the domain of the supervolcano, Toba. There are an estimated 550 volcanoes that have erupted in historic times that are considered active. There are also an approximately equivalent number of volcanoes that have erupted in the past 10,000 years, but not in historic times, that are considered dormant. The vast majority of these active and dormant volcanoes have occurred along the Earth's subduction zones.

Most of the Earth's volcanism, however, goes unnoticed by people because it occurs at the mid-ocean spreading centers, which, except for occasional islands such as Iceland, are all underwater. This global system is the largest single volcanic feature on the Earth, encircling it like the seams of a baseball for more than 35,000 miles. Here the Earth's crust is spreading, creating new ocean floor and literally renewing the surface of our planet. Older crust is recycled back into the mantle at subduction zones. The mid-ocean ridge consists of thousands of individual volcanoes or volcanic ridge segments that periodically erupt. While extremely active, there are no known supervolcanoes on the mid-ocean ridge system or in the deep ocean basins. The reason for this is the relatively thin oceanic crust and the basaltic composition of the crust, as opposed to the thicker continental settings and rhyolitic compositions for the land-based volcanoes.

BIGGER VS. SMALLER VOLCANOES

Not surprisingly, small eruptions are more frequent than larger eruptions. It just takes a longer period of time to buildup the necessary gas pressures and magma needed to accommodate a larger eruption. Estimates of the global frequency of small eruptions, producing less than 0.0024 cubic miles of volcanic material, is once every few months, whereas the frequency of very large eruptions, producing thousands of cubic miles of material, is about once every 50,000 to 100,000 years. Not as often as a great earthquake (M > 8) or giant tsunamis, but much more frequent than a comet or asteroid impact of equal threat potential. And like the comet or asteroid impact, supervolcanoes that blow their tops have the power to change the earth and civilization in ways that not even the largest earthquakes or hurricanes can. The environmental, ecological, and cultural effects of the supervolcano, Toba, will be described in Chapters 5, 6, and 8.

Caldera-forming eruptions are the biggest on Earth. The Fish Canyon Tuff in the La Garita Caldera in Southwest Colorado erupted about 28 million years ago, producing an estimated 1,200 cubic miles of magma (Mt. St. Helens, by comparison, was only 0.3 cubic miles), enough to cover a state the size of California in a layer several inches thick. This event has become known as the largest supereruption discovered to date.

Supervolcano calderas can be very difficult to pinpoint for a variety of reasons. Many form the basins of some of the most beautiful and innocent looking lakes in the world. Other are located in remote and extremely difficult places to get to (for example, Andes mountain range). Most exist where the Earth's tectonic plates collide, or where hot magma and gasses have a source for rising up from deep within the earth, usually within or near a continent. Source vents today often exist dangerously close to areas of dense human population.

Supervolcanoes are usually long-lived, that is active for periods of millions of years. And just as they can be active for such lengths of time, they can also lay dormant for the same long periods. Similar to earthquakes, the more frequent the eruption at a particular volcano, the less intense each eruption will be. Thus, the bigger a supervolcano is, the less frequently we can expect it to blow, much like larger magnitude quakes along a specific fault line, which need time to accumulate the huge amounts of stress capable of raising the roof off the Richter scale (actually saturating the scale and requiring a different measure of magnitude: the moment magnitude) and producing a great earthquake (that is, magnitude in the range of 9.0 to 9.5).

There are between 45 and 50 known supervolcanoes around the world. Many of the world's supervolcanoes are extinct, though, and others lay dormant, with barely a rumble over hundreds to thousands of years. Of those that are considered active, the fairly recent supereruption is the subject of several chapters in this book: the Toba caldera in Sumatra more than 70,000 years ago.

Supereruptions are so much bigger and more destructive than your ordinary volcanic event, that they have the potential to wipe out swaths of human, animal, and plant life, both due to initial effects (lava, pyroclastic flow) and the more damaging long-term effects on climate and the environment, resulting from ashfall and volcanic gases. Reporter Robert Roy Britt of Space.com puts it succinctly in the article "Super Volcanoes: Satellites Eye Deadly Hot Spots," when he stated, "…the eruption of Mt. St. Helens in 1980 was a volcanic sneeze compared to what scientists say America will experience one day." Not to mention the rest of the world. Gazundheit!

While the vast majority of the world's supervolcanoes have occurred along the Ring of Fire, other possible candidates exist in the area around Kos and Nisyros in the Aegean Sea, and a new volcano was just discovered 25 miles off the southern coast of Sicily, the base of which covers an area larger than Rome, and only about 60 miles south of Europe's largest active volcano, Mount Etna. Some 35,000 years ago, a massive eruption occurred in the Plegrean Fields west of Naples, Italy. While not a supervolcano, ejecting only 20 cubic miles of material, the eruption unleashed enough ash to cover large areas of southern Europe, a place now populated with millions of people.

Supervolcanoes associated with continental mantle hot spots include three Yellowstone calderas, which comprise Yellowstone Park located in Wyoming.

The Yellowstone supervolcano is one of the most massive supervolcanoes in existence. Some scientists speculate that the force of a Yellowstone eruption would equal 1,000 Hiroshima bombs exploding *every second*. The most recent supereruption at Yellowstone formed a caldera more than 50 miles long and 27 miles wide.

Another potentially devastating supervolcano is found in Long Valley in east central California close to major resort and skiing areas. While the supereruption that formed the Long Valley caldera occurred about 760,000 years ago, ongoing signs of unrest in and near the caldera serve as a reminder to people of the potential threat of an eruption.

And should you ever visit the gorgeous Lake Taupo in the North Island of New Zealand, don't let the serene setting fool you. This was the site of a massive magnitude 8 (VEI) supereruption some 26,500 years ago, large enough to take down the entire northern part of the lake basin. Taupo is still active, but its next eruption isn't expected to be as big as the original caldera-forming eruption.

One very chilling aspect of supervolcanoes is the sheer amount of force behind the eruption. According to the Geological Society of London, intensities up to several thousand cubic feet of magma per second have occurred during supervolcanoes. As a means of comparison, they point to the Soufriere Hills, Montserrat volcano. During five years of eruptions, the entire volume of magma that volcano produced is equal to the amount discharged by a supervolcano *in just a few minutes.*

As stated previously, the largest supereruption on earth within the past two million, and possibly the last 28 million, years is the subject of this book. Toba's eruption caused global temperatures to decrease by 3 to 9°F, and perhaps by as much as 18°F during growing seasons in middle to high latitudes, according to New York University Geologist Michael Rampino, wiping out between 80 and 90 percent of human life, and as much as three-quarters of all plant life in the Northern Hemisphere.

In addition to causing a sociological and environmental catastrophe of the worst dimension, Toba also changed the course of human evolutionary history, as we shall see in Part Two of this book.

CHAPTER 2

Wrath of the Gods: Volcanoes in Myth and Religion

The Earth is a storyteller, a maker of myths. She tells her tales in the form of natural events that shape the mountains and carve out oceans and valleys, shifting and changing the landscape below and sky above. From the dawn of humanity, we have been recording those events in the form of creation stories, legends, myths, and even religious beliefs that echo the chaotic upheaval witnessed by those who walked the restless Earth.

Early humans interpreted the activity of the world around them on wall paintings and cave art, and through oral stories passed on from generation to generation. Later, the written word allowed for a deeper expression of the Earth's cycles of birth, death, creation, and destruction. Imagine living alongside volcanoes, watching the mighty hills and deep calderas smoke and churn out lava, and rumble and growl and erupt and explode even as your tribe hunted, gathered, mated, and killed to survive. What awe it must have inspired. What fear it must have instilled.

Without a scientific understanding of nature, early humans instead adopted volcanoes into the fabric of their physical, spiritual, religious, and emotional behaviors and beliefs, creating a way of living devoted to interpreting, and cooperating with, the actions and motives of angry gods and goddesses imposing their wrath upon mortals. If they could not correctly interpret those actions and cooperate to the satisfaction of the deities, they would perish from fire, flood, famine, eruption, earthquake, and a host of other purely natural disasters, which, to those who came before us, were punishments for some uncertain sin.

THE MYTHOLOGY OF VOLCANOES

Naturally, people who lived closest to volcanoes were the most influenced by them. According to noted mythologist Mircea Eliade in *The Universal Myths: Heroes, Gods, Tricksters and Others*, flood stories are predominant, but all types of cosmic cataclysms are widespread among primitive peoples. These stories usually focus on the small number of survivors (sometimes just one single couple such as Adam and Eve of the Hebrew Old Testament), and view the cataclysm as an end-of-the-world punishment by angry gods (or one God) because of human misbehavior. Some of these stories speak of the cyclical nature of destruction, followed by a period of rebirth and renewal.

A thousand-year-old Toltec Indian creation/destruction myth tells of the creation of the five worlds and five suns. But, because of human actions, the first world and first sun were destroyed. The second sun was destroyed because of human lack of wisdom. The third sun, the sun of fire, sent volcanic eruptions, earthquakes, and fire to destroy the third world, again because the people were not properly making sacrifices to their gods. Always, humans were to blame for acts of nature not yet understood. This Toltec myth, called "The Scabby One Lights the Sky," continues with the fourth world destroyed. Then the Gods get together before the fifth world is about to be destroyed and choose among them one who will "jump into the fire" at the top of a pyramid. This obviously speaks of a volcano, which the unfortunate God, Tecciztecatl, was to leap into as a sacrifice, and after days of purification rites, he attempted to do so four times, but each time was driven back by his own palpable fear of the fiery heat.

That's when the lowly God Nanautzin stepped in. Known as "the scabby one" for his ugly, scab-covered body, Nanautzin hurled himself into the "pyramid of fire," and his sacrifice redeemed the Gods and returned light to their fifth world.

Ancient Toltec stories talk of a "blood rain from the sky" that fell in the form of a long, heavy rain of "flaming firestones and blood," falling on homes and bursting into flames, consuming crops and forests, and even the clothing of people who tried to escape the fiery fury. This story, found in a 1558 Spanish manuscript, goes on to say that the Creator was cleansing the Earth to bring about an end to one age, and the rebirth of a new age. Strikingly, those who survived were forced to deal with thick, dark clouds covering their land for two decades—very similar to the ash cloud that blocks the sun for years following a supereruption.

Even Judeo-Christian eschatology mentions a potentially volcano-related catastrophe as a means of bringing about a new creation: a new Earth. Saint Ephraem Syrus speaks of an end-time where "the heaven and earth shall be dissolved, and darkness and smoke shall prevail. Against the earth the Lord shall send fire, which lasting 40 days shall cleanse it from wickedness and the stains of sin." What Syrus may have been imagining mirrors the eruption of a volcano, and the subsequent darkening of the atmosphere due to the ash flow and gasses. Forty days of fire can be attributed to the ongoing magma flow over the Earth. This concept of cleansing often accompanies stories of fire and destruction.

Volcanism may also be linked to one of the most intriguing stories of the Hebrew Old Testament: the exodus of the Israelites out of Egypt. In a December 1985 *New York Times* article, John Noble Wilford reported on the discovery of tiny glass fragments from a volcano that lent support to a theory linking the miracles witnessed during the exodus with the eruption of Thera, a Greek island in the Santorini archipelago, approximately 3,600 years ago. This island is associated with the legend of the sunken lost continent of Atlantis, but may also be responsible for the "deep darkness over the whole land of Egypt" reported in the book of Exodus, as well as the ensuing tsunami caused by the eruption that may have created the "parting of the waves" of the Red Sea (probably the Sea of Reeds—see Chapter 3) that offered safe passage for the Israelites, and certain death for the pursuing Egyptians.

A Hopi prophecy, according to *American Indian Myths and Legends*, by Richard Erdoes and Alfonso Ortiz, states that the end times will come with a final warning of earthquakes, volcanoes, and eclipses, and that if the warnings are not heeded, the world will be destroyed, and a new one will take its place.

American Indians have many myths involving volcanic activity, no doubt due to their actual experiences with the mighty volcanoes of the western lands. A Yakima creation story states that the Great Chief Above made the world by throwing mud up from shallow waters to create mountains. Then, in clear Adam and Eve style, he made the natural world and a man out of a ball of mud to tend to it. The man was lonely, so the Great Chief Above made a woman. This man and woman then proceeded to anger Mother Earth, and she "shook the mountains so hard" that the falling rocks killed much animal life and dammed up streams, creating waterfalls and rapids. The Yakima believed the Great Chief Above would one day again overturn the mountains, and that the spirits of the dead could be heard inside the mountains wailing in response to their living mourners below.

The evolution and migration of the Cheyenne people includes a tale of the beginning of Great Medicine's creation of the Earth and the heavens. There were three kinds of people, including the red men who had long hair. The red people seemed to be the Great Medicine's preferred people, and he taught them to band together and live off the land, away from the hairy people and the white people Great Medicine had also placed upon the Earth.

Eventually, the red people moved south and, for hundreds of years, lived with volcanic activity. "The earth shook, and the high hills sent forth fire and smoke," says one Cheyenne tale, "Great Medicine Makes a Beautiful Country," which interestingly tells how the red people adapted to the aftereffects of an eruption. "During the winter there were great floods. The people had to dress in furs and live in caves, for the winter was long and cold...." The long volcanic winter, no doubt, turned the sky dark with ash, blocked the life-giving warmth of the sun's rays, and destroyed much of the water and food sources.

Global cooling, with falling ash and clouded skies, following a massive volcanic eruption, may also account for a Slavey native myth called "Keeping

Warmth In A Bag." In the beginning, according to this legend, there was a long winter and the sky was clouded and dark, hiding the sun. It snowed continuously for three years, until all the animals got together and held council to decide what to do. This abnormally long winter resulted in the animals blaming the bears, which had not been seen in three years, for the lack of heat and warmth. The story goes on to tell of how the animals tricked some young bears into giving them a bag of heat, which then melted the snows and warmed the Earth.

One of the most obvious volcano myths of the Native Americans is the creation story of the Modoc, called "When Grizzlies Walked Upright," in which the Chief of the Sky Spirits gets tired of living in the Above World and decides to carve a hole in the sky and push the snow and ice down to make a great mountain—Mount Shasta.

Sky Spirit then walked down the mountain, peppering it with trees with the touch of his fingertip, and with his walking stick he broke off pieces to create beaver, otter, and fish. Soon, Sky Spirit decided to bring his family to live in this new creation, and "he made a big fire in the center of the mountain and a hole in the top so that the smoke and sparks could fly out." Each time he put a log on the fire, the mountain would tremble and sparks would fly. Little do the people living near Mount Shasta today realize that the volcanic rumblings of their beloved mountain are to be blamed upon Sky Spirit putting another log on the fire!

But origin myths describing volcanic or post-volcanic activity are not limited to the Native American continent. The Greek origin myth lists the first three creatures of Mother Earth and Father Heaven who possessed the strength of Earthquake, Hurricane, and Volcano, and includes a telltale story of the destruction of the Titans in a fiery holocaust at the hands of Father Zeus. "Then Zeus no longer held back his power…From Heaven and Olympus he swings, tossing his lightning-darts. The bolts he flung were fierce with thunder and lightning, and thickly they came out of his strong right hand with a sacred flame rolling: the life-giving earth with a shudder of sound took fire, and measureless forest crackled around. All land was seething and heaving…Lapped round with a fiery stream stood the earthborn Titans. Numberless flames were blown to the brightening aether…" And it seems the son of Zeus was even more inextricably linked to volcanoes. Hephaestus, also known in Roman

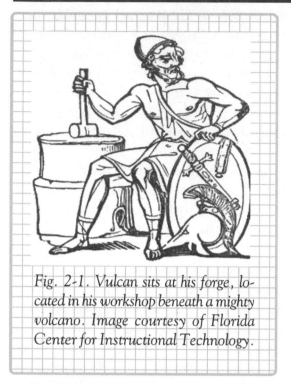

Fig. 2-1. Vulcan sits at his forge, located in his workshop beneath a mighty volcano. Image courtesy of Florida Center for Instructional Technology.

mythology as Vulcan, was called the God of Fire, and both Roman and Greek myth associate him with a workshop underneath a volcano, which erupts when he forges weapons and tools. His Roman name Vulcan means "volcano." The word *volcano* comes from the little island of Vulcano in the Mediterranean Sea off Sicily where, centuries ago, the people living there believed that Vulcano was the chimney of Vulcan's forge. The "smoke" from his forge may have been volcanic ash and pyroclastic flow sent skyward from an eruption.

Interestingly, Vulcan is considered a kindly, peaceful god, popular with both mortals and immortals, and a patron of the handicrafts and protector of the smiths.

VOLCANO GODS AND GODDESSES

✳ **Aetna:** The daughter of Uranus and Gaia, she is the personification of Mt. Etna, the volcano in Sicily. Eruptions of Mt. Etna are attributed to the presence of the giant Typhon, which lies buried beneath the volcano.

✳ **Typhon:** The offspring of Gaia and Tartarus. Described as a beast with a hundred Typhon's horrible heads, and lava and red-hot stones pouring from his gaping mouths. Zeus battled Typhon (and had the advantage), and as Typhon tore up Mt. Aetna to hurl at the gods, Zeus struck it with a hundred well-aimed thunderbolts and the mountain fell back, pinning Typhon underneath the volcano, belching fire, lava, and smoke through the top of the mountain. The word *typhoon* is said to come from Typhon.

* **Cherufe:** A gigantic lava creature in Chilean myth that lives inside volcanoes. Cherufe feeds upon young virgins. The Sun God sent two of his warrior daughters to guard Cherufe to protect the local people.

* **Masaya:** Nicaragua's goddess of volcanoes. Masaya is the most active volcano in the region and its eruption in 4550 B.C. is considered one of the largest in the last 10,000 years. Masaya is known for its large releases of sulfur dioxide. In 1981 alone, more than 500,000 tons were released from the Masaya Volcano. Two years prior, Masaya was designated a national park (Parque Nacional Volcan Masaya).

* **Mephitis:** A Roman Goddess who protected people in volcanic regions from poisonous gasses and damage from volcanic vents.

* **Llao and Skell:** Native American Klamaths living near Mount Mazama, a pre-Crater Lake volcano, some 6,000 years ago believed these two warring Gods were responsible for fiery avalanches coming down from the mountain. Llao was Chief of the Below World, and he fought with Skell, Chief of Above World, hurling rocks and flames. Llao was injured and fell, becoming the huge crater, which would later become Crater Lake. Volcanologists now know that the caldera was formed from large explosions more than 7,000 years ago.

* **Louwala-Clough/Loowit, Klickitat/Pahto, and Wy'east:** Northwest American Indians living near Mount St. Helens, which the Indians called "Louwala-Clough" or "smoking mountain," believed that the mountain was once a lovely maiden named Loowit. Two sons of the Great Spirit Sahale fell for her but she could not choose between them. The two sons, Klickitat and Wy'east, fought over her, causing destruction, and Sahale angrily smote the lovers and erected a mighty mountain where

each fell. Loowit became Mount St. Helens, Klickitat became Mt. Adams, and Wy'East became Mt. Hood.

✳ **Takhoma/Tahome/Ta-co-bet:** Mount Rainier was well known to the Native Americans, who knew the mighty volcanic mountain as "Takhoma" for "big mountain," "Tahoma" for "snowy peak," and "Ta-co-bet" for "place where waters begin."

✳ **Popcatepetl and Iztaccihuatl:** Two mighty adjacent volcanoes in the Valley of Mexico, once said to be lovers who could not bear to be out of each other's sight, according to Aztec legend.

PELE AND OTHER HAWAIIAN MYTHS

One of the most well-known volcano divinities is Pele, the Hawaiian (Polynesian) Goddess of Fire who lives in the volcano Kilauea on Hawaii's south side. One of the most active volcanoes on Earth, it is home to the powerful and tempestuous beauty who caused eruptions during her moments of rage and anger. Legends say, Pele caused volcanoes to erupt by digging with her *Pa'oe*, or magic stick, and that the Pig God Kamapuaa once made war with her after trying to win her love. For centuries, Hawaiian islanders would toss live pigs into Kilauea's crater to try to appease the angry Pele and keep her from boiling over. We devote more space to Pele because of the relationship of her legend to the modern hotspot theory of the formation of the Hawaiian Islands, and because of her "curse" that affects people's lives to this very day.

Pele was born in Tahiti, but exiled by her father, Kane Milohai, because of her bad temper, which, along with her penchant for jealous outbursts, is responsible for her high place in Hawaiian myth as the most powerful of Goddesses. Even the tear-shaped lava droplets around the Kilauea Volcano are said to be her tears, and her hair is the volcanic glass that takes on the appearance of fine strands. Pele is said to have sailed from Tahiti in a canoe guided by her brother, the Shark God Ka-moho-ali'i.

There is an excellent discussion of the interplay of Hawaiian legend and Earth science in the book by Simon Winchester, *Krakatoa: The Day the*

World Exploded August 27, 1883. The ancient Hawaiians had long suspected differences in ages as one proceeds southeast from the relatively old, volcanically quiet island of Kauai to the young, bubbling, fiery Big Island of Hawaii. The seafaring islanders based their suspicions on observations of differences in erosion and vegetation between the islands. This age difference was incorporated in the legend of the mythical flight of Pele from Kauai to Hawaii, with stops at the intervening islands of Oahu and Maui. The legend handed down from generation to generation alludes to an eternal struggle between Pele and her older sister Na-maka-o-kaha'i, the Goddess of the Sea. Each time Pele landed on a new island, she created a new volcano and her sister then flooded the area out, until they went head to head in an epic battle near Hana, Maui where Pele was torn apart. Upon her death, she became a true god and made Mount Kilauea on Hawaii her current home. It is known as the Navel of the World.

Pele is often portrayed in art as a woman of fire or with flames surrounding her. Although she is considered a protector of Hawaiian people, she is said to curse nonnatives who return to their homelands with native rock in hand. Each year, thousands of lava rock pieces are shipped back to Hawaii from tourists who claim to have experienced the curse of Pele, accompanied by notes and letters begging her for forgiveness! An interesting example of the curse of Pele was reported in the *Los Angeles Times* on May 17, 2001, by reporter Julie Cart. The report concerns a Timothy Murray and his incredible run of misfortune.

Timothy Murray had a comfortable life: a college education, good jobs, fulfilling relationships. To quote Murray, "I've always had really good luck."

But that was before he crossed paths with madam Pele!

Murray's luck went south in 1997 after he went to Hawaii to accept a new job. When the job fell through, Murray consoled himself with a trip to the Big Island and Hawaii Volcanoes National Park. Entranced by the Island's black sand beaches, Murray did what tourists often do. He took home a memento, scooping up the sand in a soda pop bottle.

When he returned home to Port St. Lucie, Florida, Murray's good fortune had fled. His beloved pet died. The five-year relationship with the woman he was to marry fell apart. He began to drink heavily. Finally, FBI agents, who had been tracking him from Hawaii, arrested him in a computer copyright infringement case.

"My life literally fell apart," Murray, 32, says of the three years after he scooped up the sand on that beach in Hawaii. "One minute you're working and you're law-abiding and you've got money in the bank. The next minute you are sitting in a federal penitentiary in Miami. I couldn't figure out what was happening or why. Even the FBI agents said they never arrest people for what I did. They told me, 'You really must have pissed someone off.' After some research, I figured out who it was."

Pele!

Following that revelation, Murray mailed the sand back to Hawaii and included a prayer to Pele for forgiveness.

For those willing to tempt fate, take heart that according to some reports the curse was originated by a park ranger in 1946 to discourage visitors from taking samples of volcanic rocks as souvenirs. The authors, however, would choose not to fool with mother Pele.

Supposedly, one of the ways to stay on Pele's good side is to make an offering to her lava of the finest gin you can find. A bottle of Tanguery or Bombay Gin should suffice provided, of course, that one hasn't sampled the offering. This smacks of something W.C. Fields would have concocted if he had been director of tourism for the Big Island and in charge of personally handling all offerings.

LOIHI

The migration of volcanic activity from Kauai to Hawaii described in the legend of the ancient Hawaiians has been confirmed by modern geologic and radiometric-age-dating studies. As we saw in Chapter 1, as the Pacific Plate continues to move west-northwest, the island of Hawaii will eventually move beyond the reach of the hotspot and surrender its bubbling, fiery personality to the budding submarine volcano Loihi, presently forming about 20 miles off the southern coast of Hawaii. Loihi has already risen from the depth of the ocean floor to within approximately 0.6 miles of the ocean surface. Assuming Loihi continues to grow, it will become the next island in the Hawaiian chain, or merge with the "Big Island" of Hawaii. Either way, Pele may once again be on the move to more fiery confines.

A recent Loihi myth has already been reported that involves a queen, who, at one time, resided on the Big Island, and a hungry, stomach-rumbling sea monster: The Loihi seamount is the hungry sea monster, and ongoing earthquake activity is evidence of its rumbling stomach. According to the myth, the queen made a deal with the sea monster that if he stayed away from the islands and did not eat any more Hawaiian people, she would feed him pineapples every day. Things went along just fine until one day the queen passed away, and the supply of pineapples ceased. The sea monster still sits in his lair in the ocean off the southern coast of the Big Island and waits for his pineapples, his hunger and anger building. People claim that they can occasionally hear the monster's stomach rumble.

Since 1970, when Loihi suddenly became active, there have been more than 10 earthquake swarms (stomach rumblings) originating in or near the Loihi seamount (sea monster). One of the largest swarms occurred in July and August of 1996, and consisted of more than 4,000 earthquakes, the largest a magnitude 5. The continuing activity is evidence of the growth of Loihi, and has led some people to the belief that one day the sea monster will not be able to tolerate his hunger and will emerge from the depths of the ocean to feed on pineapples, or Hawaiians, if pineapples are not available. Hawaiians and tourists, who might be mistaken for Hawaiians, need not worry or stock up on pineapples because scientists estimate that the sea monster, Loihi, will not emerge for at least another 50,000 to 100,000 years.

NORSE MYTHOLOGY

In Norse mythology, the Mountain Giants were thought to be especially nasty. They, along with the Frost Giants, represented the brutal powers of Earth, battling with the heavenly divine powers for dominion. The Norse have their own version of the Christian Bible's Book of Revelations called the Elder Edda, which includes a prophecy of an end time with clearly volcanic implications.

"The sun turns black, earth sinks in the sea, the hot stars leap from the sky, and fire leaps high about heavens itself..." but this catastrophic event was to be followed, as in other creation myths, by a new heaven and Earth under the reign of One who was above even the high god Odin.

HINDU TALES

Unlike the angry and jealous gods and goddesses who often set volcanoes into motion, the Hindu tales of the divine Indra attribute him with more restorative powers. According to the Rig Veda, Indra "made fast the tottering earth…made still the quaking mountains." Similar to Zeus, he was the possessor of mighty thunderbolts, but Indra chose to use his to protect the people. The Rig Veda goes on to say, "even the sky and the earth bow low before him, and the mountains are terrified of his hot breath." Not every primitive culture or early civilization thought of nature as uncontrollable, and many propped their gods and goddesses above the highest tops of the mightiest volcanoes, ready to silence the roar of a pending eruption. Not that these gods and goddesses could stop eruptions, as geological history has proven, but in the minds of the people dealing with such devastation, the influence of the divine ones surely must have been felt once the eruption subsided, and some semblance of normalcy could at least be hoped for.

RUSSIAN MYTHOLOGY

One of the most volcanically active regions in the world is Russia's Kamchatka Peninsula. This area, which is part of the notorious Ring of Fire, has the highest density of volcanic activity in the world, and is often referred to as Russia's Yellowstone. The original inhabitants of the peninsula are native peoples known as Koryaks, Itelmens, Chukchis, and Tunguses. The Koryaks were most populous and their creation myth speaks of their relationship to the nearby volcanoes that cast shadows upon the fertile valleys they called home. The Koryak creation myth is actually quite romantic, with Great Raven Kutkh creating Kamchatka from a dropped feather. Kutkh then created man to inhabit the new world, and of course a woman duly followed. Only, the men all fell in love with the same beautiful woman, and as each died, he became a mountain. As their hearts burnt with fiery love for the first woman, they became volcanoes, the eruptions symbolic of the rising passion of unrequited love.

Even lesser tribal peoples of the region, including the most ancient Northern Itelmens, adopted volcanoes into their creation tales, but to them, the volcanoes were the dwelling places of demons. When a volcano erupted, it was because the mountain demons, or kamuli, who lived on fish, did their fishing at night, and the light of the fiery eruption is the cooking fire of the kamuli preparing their fish to be eaten. Itelmens believed they could gain the favor of the kamuli by tossing meat into the volcanoes to keep them from erupting.

ASIAN MYTHOLOGY

Another active zone along the Ring of Fire lies near Japan. Earthquakes, volcanoes, and typhoons have been a part of the ancient Japanese culture since humans walked the earth. Ancient Japanese believed that the Gods created Japan as a special place, and that lesser gods called Kami were sent to preside over the mountains, forests, winds, and oceans. The respect and awe of the Japanese people for the natural world with which they interacted is reflected in the Japanese religion, Shinto, which teaches that nature is alive with spirits. Shinto means "the way of the gods," and borrows concepts from Buddhism.

Japan is a country of mountainous terrain, so it is easy to see how volcanoes can be such a large part of their folklore and beliefs. With 109 volcanoes of varying states of activity, ancient and modern Japanese have incorporated the presence of volcanoes into their everyday lives. Mount Fuji is especially important to followers of Shinto. It is an exceptionally symmetrical and active stratovolcano, one of the most common types of volcanoes found along Ring of Fire. Japanese legend tells of the creation of Mount Fuji at the hands of a woodsman named Visu, who awoke one night to a very loud underground noise. Thinking it an earthquake, he grabbed his family, but when he went to the doorway of his home to escape, he found that his flat land was now a mountain. So in awe was he of this now beautiful mountain his home sat upon, he named it "Fuji-yama," the Never-Dying Mountain.

Shinto pilgrims have been making regular treks to the top of Mount Fuji, known to the native Japanese as Fujisan, for hundreds of years to pray for peace and prosperity. Legend has it that the Goddess of Fuji, Sengen, would throw the impure down from the mountain. But that hasn't stopped upwards of 500,000 people each summer from making the journey. Women were once excluded, but can now make the pilgrimage to the top of the volcano as well. And as far as we can tell, there have been no reports of bodies being mysteriously tossed from the mountain by unseen forces!

Indonesian legend also mentions the creation of the Tengger Crater in Java, which was dug out with half a coconut shell by an ogre in love with a beautiful princess. He had only one night to do so, according to the deal he had been given by the King, but when the King saw that the ogre just might complete the task, he sabotaged the ogre by ordering his servants to pound rice all night, which, in turn, woke the cocks who then started crowing early. Thinking it sunrise, the ogre believed he had failed and lost the deal, so he tossed away the coconut, which became Gunung Batok, and then, defeated and broken, died of exhaustion.

Island nations often have rich volcanic history, and the Maoris of New Zealand are no exception. These Polynesian peoples have lived on New Zealand since the 14th century, and their lore is rich with tales centering on the Ngauruhoe, Tongariro, and White Island volcanoes. In one such tale, medicine man Ngatoro is climbing the mountain Tongariro with a woman called Auruhoe. Ngatoro warned his followers not to eat while he was gone, so that he would have their extra strength on the cold mountain, but when he did not return for some time, they of course disobeyed and broke their fast (sound similar to Moses and the golden calf story?). When Ngatoro and Auruhoe got terribly cold, they prayed to his sisters in Hawaiki, who then called upon fire demons to save Ngatoro. The fire demons swam underwater and emerged at White Island, and the land burst into flames. The demons then continued underwater and burst through the summit of the mountain, creating the volcano Ngauruhoe, which warmed Ngatoro, but caused the woman Auruhoe to perish. Ngatoro then threw the woman's body into the volcano.

In other myths, the mighty volcanoes of the New Zealand islands were thought to be giants battling for the love of a beautiful woman, or, as one legend associated with the 1886 eruption of Mount Tarawera states, the eruption was said to be punishment of the people of the Te Ariki village for eating forbidden honey.

ICELANDIC MYTHOLOGY

Further north, in the land of ice and snow, millions of years of lava from the Mid-Atlantic Ridge piled up and created Iceland, a country rich with modern volcano myths, many of which centered on the fiery antics of Hekla. Hekla is the most active, and most famous volcano, in Iceland. Located in the south-central region, it consists of an elongated stratovolcano, which lies along a major northeast/southwest trending fissure system. Hekla has been erupting regularly since A.D. 1104, at intervals of 10 to 100 years. It last erupted in 2000.

Christians living in Europe thought Hekla (Benjamin Franklin's "Hecla" in Chapter 3) was the doorway to Hell or Hades and believed the flying lava and rock projectiles were the spirits of the dead, hissing in pain and agony as they flew through the sky (the sound no doubt due to the generation of steam as the cooler, moist air comes in contact with the red hot lava). Legends of Hekla suggest it was a meeting place for black Witchcraft and dark magic.

Hekla remains volcanically active, tossing fire fountains into the sky from time to time, and spewing lava down its sides, but modern humans now understand it to be nothing more than a show of natural force and not the wrath of demons, or the punishment of angry gods, that our ancestors once believed it to be.

ALASKAN MYTHOLOGY

The Eskimos of Alaska continued to incorporate volcanoes, which are plentiful in their region, into their folklore as recently as 1783, when the eruption of the Icelandic volcano Skaptar Jokull inspired the Legends of Old Willie. According to *People of the Kauwerak: Legends of the Northern Eskimo*, by Laurel L. Brand, these stories were first recorded by a man named William A. Oquilluk, known as "Old Willie." One story focuses on the aftermath of an eruption, and begins with the summer that never was, when ash clouds and sulfur from the eruption blew toward Alaska just as hunting season was about to start. Of the villagers, only 10 survived the "volcanic winter." Two of the survivors were a grandmother and granddaughter who had no one to hunt for food for them. Luckily, they had been smart enough to store some of the earlier summer foods, and they went outside, with the cold, dark cloud hanging over them, to find the other villagers dead. Living off sealskin, they survived the long winter.

A similar Old Willie legend recounts the story of two other villagers who survived the long winter, this time a mother and her small boy, by trekking more than 200 hundred miles to reach another village.

MODERN MYTHS

Obviously, locales rampant with volcanic activity are also rampant with myth, legend, and lore involving eruptions and their aftermaths, and we usually think of these stories as originating among ancient and primitive peoples with no understanding of the science behind the supernatural. But oddly enough, there are two very modern religions that incorporate volcanoes into the foundations of their belief system.

SCIENTOLOGY

Scientology is a religion steeped in controversy, to say the least, thanks in no small part to the antics of celebrity followers. Yet, few people know that the doctrine behind the religion based upon the writings and teachings of founder L. Ron Hubbard has its own creation tale, and one that implicitly involves volcanoes.

The use of the volcano as a symbol for both the book cover of *Dianetics*, which explains the science behind Scientology, and for Scientology itself, comes from the story of Xenu, an alien ruler of the "Galactic Confederacy" who brought billions of people to Earth in spaceships, deposited them around the bases of volcanoes, and then blew them up with hydrogen bombs. This traumatic and cataclysmic event is known as "Incident II" and became a critical part of the Scientology doctrine in Hubbard's 1967 story "Operating Thetan Level III" (OT III). Part of Scientology's secret "advanced technology," taught only to committed and wealthy contributing members, OT III states that some 75 million years ago, Xenu apparently decided to eliminate excess human population from his 26 stars and 76 planets by paralyzing them and loading them onto space planes bound for "Teegeeack/Earth."

Once on Earth, the people were literally stacked around the bases of existing volcanoes across the globe, and hydrogen bombs were dropped into the volcanoes and simultaneously detonated. As Hubbard's own script for the unmade film *Revolt in the Stars* states:

"Simultaneously, the planted charges erupted. Atomic blasts ballooned from the craters of Loa, Vesuvius, Shasta, Washington, Fujiyama, Etna, and many, many others. Arching higher and higher, up and outwards, towering clouds mushroomed, shot through with flashes of flame, waste, and fission. Great winds raced simultaneously across the face of Earth, spreading tales of destruction...."

Some critics have pointed out that many of the volcanoes Hubbard claims Xenu blew up did not exist 75 million years ago, and that other regions have no volcanic history whatsoever, but as with the case of many, if not all, religions, the facts often get shoved behind the fiction.

Of all the most interesting and creative ways volcanoes have influenced myth, folklore, legend, and religious belief, though, the hands-down winner

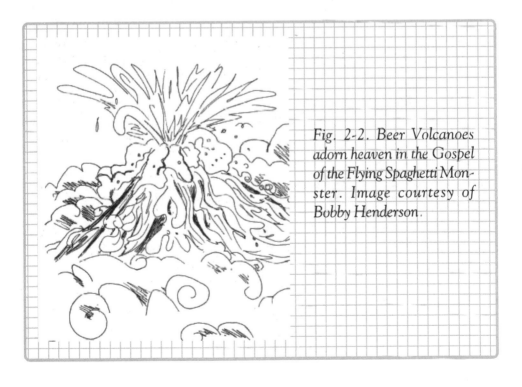

Fig. 2-2. Beer Volcanoes adorn heaven in the Gospel of the Flying Spaghetti Monster. Image courtesy of Bobby Henderson.

goes to a fairly new religion that has made volcanoes a revered part of their heavenly doctrine. The Church of the Flying Spaghetti Monster was created in 2005 by computer scientist Bobby Henderson as a rather clever protest against the Kansas State Board of Education's decision to allow Creationism to be taught in their schools.

Henderson wrote a letter to the school board and started a Website to launch his own version of "intelligent design" as a direct way of poking serious fun at the Kansas decision. He created his own religion that worships a Flying Spaghetti Monster (FSM), also known as His Noodly Appendage, and a whole doctrine of belief known as Pastafarianism. Henderson's parody may sound silly, and you have to see the Website to get the full Noodly effect, but his goal was to present a serious case for why intelligent design simply would not float.

The Church of the Flying Spaghetti Monster now boasts thousands of "followers" committed to a hilarious belief system that states evidence for evolution was planted by the FSM; that prayers must end with the concluding word "Ramen"; and that the pirates are absolute divine beings. Henderson puts forth the theory that all natural disasters are the result of the shrinking number of pirates since the 1800s, also showing that global temperatures have risen as the pirate population has decreased. Laugh, but Henderson's purpose was to prove to the Kansas school board that correlation does not equal causation!

Where do volcanoes fit into this doctrine? Well, be a good follower of the FSM and you will end up going to Heaven, a land of pure joy that is made up of two things: a stripper factory, and beer volcanoes "as far as the eye can see."

THE POWER OF VOLCANOES

To have such a deep, profound influence on the spirit, soul, and psyche of an entire species, volcanic activity must have appeared so horrific, so powerful and inescapable, that it seared itself into the memory of those experiencing it, something we will delve into more fully in a later chapter.

In the next chapter, we will examine the ecological effects of some of the largest explosive eruptions that have occurred in the historic past. In terms of what we considered in this chapter, we will look at how these violent eruptions impacted humans: the very foundation of civilization.

But to understand exactly why volcanoes have permeated our myth, religion, and folklore from the dawn of time to this very day, when we still find the foundation of our faith shattered with each new devastating natural disaster, we must get closer to, and go deeper inside, the beast itself.

CHAPTER 3

Aftereffects of Volcanoes

Our goals in this chapter are twofold. First, we want to follow the evolution in our understanding of volcanoes as it progressed from myth, speculation, and anecdotal observations to scientific observations and theory regarding the effect of volcanic activity on the global environment. At the same time we want to see how scientists, starting with what they have learned from historical volcanoes of VEI 7 or less, are attempting to apply that information to an assessment of the environmental aftermath of supervolcanoes. The extrapolation of scientific understanding of the underlying physics of the environmental impact of volcanic activity is particularly important in assessing the aftereffects of a possible future supereruption.

UNDERSTANDING VOLCANOES

Basically, observations of the atmospheric effects of volcanic eruptions throughout recorded history until the late 19th century focused on *what* rather than *why* things happened. This was in part a result of the fact that much of the scientific information that is taken for granted today was unknown at the time of the 1783, 1815, and 1883 historic volcanic eruptions. For instance, volcanic gases injected into the Earth's stratosphere can affect the global climate and the ozone layer. And yet, it wasn't until the year 1900 when an innovative program of unmanned balloon soundings of the upper atmosphere by the French physicist Teisserenc de Bort unexpectedly revealed a large region with relatively constant temperature, followed by an increase in temperature above that height. He named this region the "stratosphere." Since that time, the stratosphere has been probed repeatedly by balloons and airplanes, as well as remotely from the ground and space-based satellites.

Likewise, while the ozone was discovered in 1839 by C.F. Schonbein as a byproduct of electrical discharges, it wasn't until the early years of the 20th century that researchers discovered an ozone layer in the stratosphere. The researchers, especially an Oxford scientist named Gordon Dobson, conducted experiments showing that the stratospheric ozone layer strongly absorbed light in the ultra-violet region of the electromagnetic spectrum: a fact of special significance to Earth's flora and fauna.

There are several ways that volcanoes affect the natural and human environments. One is the impact that volcanic eruptions have on weather and climate. We can think of weather as the current and short-term state of the Earth's atmosphere in terms of temperature, precipitation, cloud cover, and wind conditions. Climate, however, can be defined as the average condition of the atmosphere near the Earth's surface at a particular location through a period of time ranging from months to years to millennia or longer, taking into account temperature, precipitation, wind speeds, and directions, and related phenomena such as frequency and intensity of storms. Volcanic activity is just one of many factors that can affect weather, climate, and the natural and human environments, and the role of volcanic activity is a subject of considerable ongoing interest. The availability of data from a diversity of ground and space-based instrumentation, and greater computing power to analyze that data, bodes well for a better understanding of the interaction of volcanic eruptions and our environment, as well as the development of improved models of climatic behavior over a range of time scales.

VOLCANOES AND THE ATMOSPHERE

Volcanoes that erupt explosively often blast huge clouds of ash and gases to different heights in the atmosphere. The height the volcanic ejecta attain is dependent on properties of the volcanic eruption and the vertical temperature structure of the atmosphere (up to and including the stratosphere). Studies to date indicate that ash and gases erupted from some of the most explosive volcanoes are generally confined to heights of 30 miles or less. This means that there are two atmospheric layers of distinctly different temperature regimes in this height range that must be considered when discussing the potential affects of volcanic activity.

The lowest part of the atmosphere, between the ground and roughly 5 miles at high latitudes, 6 to 7 miles at mid-latitudes, and 9 miles near the equator is called the troposphere, from the Greek *tropos*, meaning change or overturning. Temperature in the troposphere generally drops sharply with increasing height, falling at a more or less constant rate of about 15°F for each mile (increase in height). Because temperature drops so rapidly with height in the troposphere, the air is vertically unstable, allowing it to rise and fall and overturn rapidly. This rapid, vertical motion, called convection, very effectively mixes together all of the gases in this portion of the atmosphere, so that any gases released at the surface will quickly find their way up to great heights. In fact, their vertical progress is only impeded by a temperature barrier between the troposphere and the region above it. This barrier, known as the tropopause, is a region where the temperature stops falling with height, and plays an important confining role for relatively small volcanic eruptions (for example, VEIs <2–3 depending on the latitude of the volcano).

STRATOSPHERE

The region above the tropopause where the temperature is, first, relatively constant and then increases with height is called the stratosphere, from the Latin *stratos* for layer. The stratosphere extends from the tropopause up to around 30 miles. The reason the temperature stops falling with height and eventually starts to increase in the stratosphere is due to the presence of the ozone layer. It is here that nearly all atmospheric ozone is naturally produced and destroyed. The temperature increase in this region is a result of the very efficient absorption by ozone molecules, which consist of three oxygen atoms (O_3) of energy from the sun over a narrow band of wavelengths corresponding to ultraviolet radiation. Amazingly, for a gas that is only present in concentrations of around five to 10 molecules in every million molecules of air, the ozone layer has a profound influence on temperature, and the existence and well-being of human life on Earth. To get a better picture of how thin a protective shield the ozone layer really is, it has been estimated that if all the ozone in the atmosphere were compressed to a pressure corresponding to that at the earth's surface, the resultant layer of ozone would be only 0.1 inches thick. The impact of volcanoes on this layer so vital to humans is discussed later in this chapter.

The increase in temperature with height in the stratosphere makes this region a very stable place where air tends not to overturn vertically. In contrast with the troposphere, where vertical wind speeds are often several feet per second, in the stratosphere they are seldom more than a few inches per second. Consequently, it takes air a very long time to be transferred from the bottom to the top of the stratosphere, unless there is a thrust of gases such as that during a highly explosive volcanic eruption. This inability to mix rapidly in the vertical direction is thought to be the principal reason why chlorofluorocarbons (the substances that carry ozone-destroying chemicals into the stratosphere), take so long to reach altitudes where the sun's energy is sufficient to break them apart, and why some of them will still be there a hundred years, or more, from now.

DRY FOGS

More than 2,000 years ago, the Greek historian Plutarch and the Roman philosopher Seneca noted that the dimming of the sun was probably caused by the 44 B.C. eruption of Italy's largest volcano, Mount Etna, on the east coast of Sicily. They suggested that the resulting colder temperatures caused crops to shrivel, which, in turn, led to famine in Rome and Egypt.

No other writings on this subject were known until a paper titled "Meteorological Imaginations and Conjectures," was read before the Literary and Philosophical Society of Manchester, England, on December 22, 1784. The author of this paper was the American Statesman Benjamin Franklin, who was stationed in Paris at the time serving as the first diplomatic representative of the young United States. In his paper he wrote that during the summer of 1783, the weather was abnormally cold both in Europe and back in the United States. The ground froze early, the first snow stayed on the ground without melting, the winter was more severe than usual, and there was "a constant fog over all Europe and a great part of North America. This fog was of a permanent nature; it was dry, and the rays of the sun seemed to have little effect toward dissipating it, as they easily do a moist fog, rising from the water." Franklin went on to speculate in his paper that the cause of "this universal fog" was either smoke "proceeding from the consumption by fire of some of those great burning balls or globes which we happen to meet with in our rapid course round the sun...or whether it was the vast quantity of smoke long continuing to issue during the summer from Hecla in Iceland and that other volcano which arose out of the sea near that island, which smoke might be spread by various winds, over the northern part of the world is yet uncertain." Two hundred years later, studies pointed to eruption of the Laki fissure in Iceland that began in late spring 1783 and lasted for several months, as the most likely cause of Franklin's "dry fog."

As we would learn later in the 20th century, "dry fogs" appear in the atmosphere when large volcanic eruptions inject massive quantities of fine silicate ash and aerosol-forming sulfur gases into the troposphere and the stratosphere. Although the ash gravitationally settles out within weeks, the aerosols spread around the globe and can remain suspended in the stratosphere for years. Absorption and backscattering of solar radiation by the volcanic particles result in a haziness in the sky and a dimming of the sun and moon.

VOLCANIC ASH

Before continuing with a discussion of the historical eruptions and their environmental and ecological impact, we will take a closer look at what constitutes volcanic ash. Volcanic ash is a form of a more general air-fall material

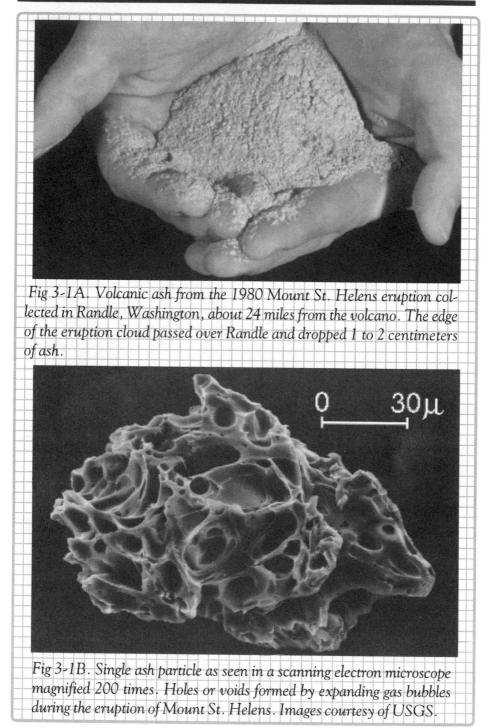

Fig 3-1A. Volcanic ash from the 1980 Mount St. Helens eruption collected in Randle, Washington, about 24 miles from the volcano. The edge of the eruption cloud passed over Randle and dropped 1 to 2 centimeters of ash.

0   30μ

Fig 3-1B. Single ash particle as seen in a scanning electron microscope magnified 200 times. Holes or voids formed by expanding gas bubbles during the eruption of Mount St. Helens. Images courtesy of USGS.

known as tephra. Tephra is typically the fragmented product of an explosive eruption and includes, in addition to ash, lapilli and volcanic bombs. The different fragments of tephra are classified according to size: bombs are the largest, defined as greater than 2.5 inches in size;. and ash the smallest, at less than 0.08 inches in size. The larger the fragment, the closer to the volcano it settles out of the eruption column. As we shall see in subsequent chapters, ash usually covers a much larger area and, as a result, can disrupt the lives of far more people and animals than the other more lethal types of volcano hazards. The actual thickness of ashfall at a particular location is dependent on the magnitude of the volcanic eruption and the strength and direction of the prevailing winds at the height of the eruption column.

Volcanic ash should not be confused with soot or ash from burning wood, leaves, or paper. It consists of rock and small volcanic glass fragments with relatively sharp edges that, as we will see in Chapter 9, can result in severe damage to lungs if inhaled. It is also extremely abrasive, mildly corrosive, and electrically conductive, especially when wet. The fact that it is electrically conductive plays an important role in the dating of historic and prehistoric explosive eruptions based on ice cores from Greenland and Antarctica.

VOLCANOES OF PAST AND PRESENT

We now turn to historical volcanic eruptions and what they have taught us that we can apply to supervolcanic eruptions. On June 8, 1783, the earth cracked open and lava erupted explosively out of a fissure some 20 miles long. This volcanic eruption was known as Laki, and exploded over and over in the following months. These initial explosive eruptions were thought to be the result of interactions of ground water and the rising basalt magma. Gases emitted by Laki were estimated to reach altitudes of about 9 miles and aerosols built up in the stratosphere caused a cooling effect in the Northern Hemisphere of neary 2°F.

Through time, the eruptions became less explosive with the eruptive style changing from highly explosive Plinian to less explosive Strombolian, and finally to Hawaiian with high rates of lava effusion. The Laki eruption is thought to have produced the second largest lava flow ever witnessed. Some estimates run as high as enough material to build the Great Wall of China a dozen times over. The Laki eruption lasted eight months (into February 1884) during which time about 3 cubic miles of basaltic lava and some tephra erupted from the volcano. Haze from the eruption was reported from Iceland to Syria. In Iceland, the haze lead to the loss of most of the island's livestock (by eating fluorine contaminated grass), crop failure (by acid rain), and the death, by famine, of one-quarter of the population of Iceland. It is estimated that 80 mega tons (Mt) of sulfuric acid aerosol was released by the eruption (four times more than the 1982 Mexican volcano El Chichon, and 80 times more than Mount St. Helens).

The eruption produced about 4 cubic miles of basaltic lava and 0.2 cubic miles of tephra. Lava fountains were estimated to have reached 2,600–4,600 feet in height. In Scotland and Great Britain, the summer of 1783 was known as the sand-summer, due to the huge amounts of ash fallout.

HISTORIC ERUPTIONS

Several of the key historic eruptions discussed in the following pages occurred along the Indonesian island arc system depicted in the following image. Proceeding from east to west along the island arc, the volcanoes are Agung (1963), Tambora (1815), Krakatoa (spelled Krakatau on image) and, possibly, one of the largest explosive eruptions in recorded history referred to as Proto-Krakatoa in A.D. 536, which may have been responsible for the formation of the Sunda Straits and the splitting apart of Java into the present-day islands of Java and Sumatra. Indonesia contains more than 130 active volcanoes, more than any other country on Earth. They comprise the axis of the island arc system, which is generated by northeastward subduction of the Indo-Australian plate. The great majority of these volcanoes lie along the topographic crests of Java and Sumatra.

MOUNT TAMBORA

While there has not been an explosive volcanic eruption on the scale of the Toba or the Yellowstone supereruptions, catastrophic explosive eruptions

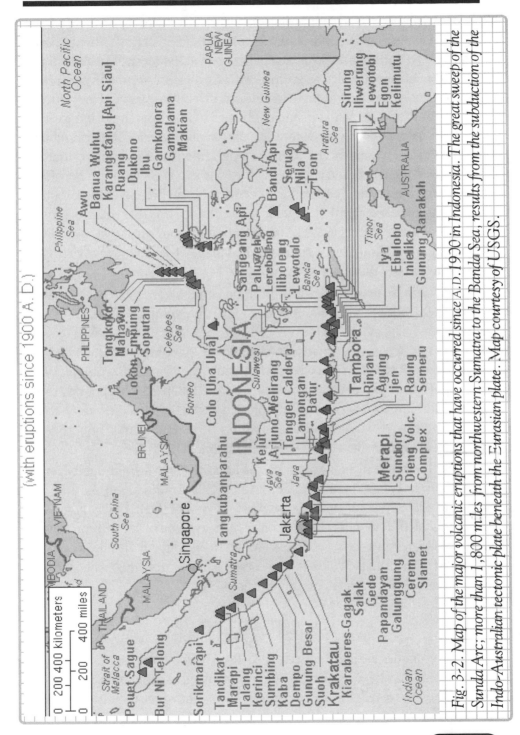

(with eruptions since 1900 A.D.)

Fig. 3-2. Map of the major volcanic eruptions that have occurred since A.D.1900 in Indonesia. The great sweep of the Sunda Arc, more than 1,800 miles from northwestern Sumatra to the Banda Sea, results from the subduction of the Indo-Australian tectonic plate beneath the Eurasian plate. Map courtesy of USGS.

have been observed, and their activity described. To grasp some idea of what's involved when a caldera forms during or just after an ash flow eruption, consider Mount Tambora, on the island of Sumbawa, Indonesia. For about three years this volcano rumbled and fumed before a moderate eruption on April 5, 1815 produced thundering explosions heard 870 miles away. The next morning volcanic ash began to fall and continued to fall, though the explosions became progressively weaker, until, on the evening of April 10 the mountain went wild. Eyewitnesses 20 miles away described three columns of flames rising from the crater and combining into one at a great height. The whole mountain seemed to be covered with flowing liquid fire. Soon these distant viewers were pelted with 8-inch pumice stones hurled from the volcano. Clouds of ash, borne by violent gaseous currents, blasted through nearby towns blowing away houses and uprooting trees. The village of Tambora was destroyed by rolling masses of incandescent, hot ash.

On April 16, booming explosions loud enough to be heard on Sumatra, up to 1,600 miles to the west, continued into the evening. Mount Tambora, still covered with clouds higher up, seemed to be in flames on its lower slopes. For a day or two, skies turned jet black and the air cold. When the eruption ended, the ash cloud drifted west and settled on all islands downwind. With the expulsion of so much magma, the mountain collapsed, unsupported from within, forming a great caldera. Lombok, 124 miles to the west, was covered by a blanket of ash 2 feet thick. Tsunamis crashed on islands hundreds of miles away. Waves and ashfalls killed more than 88,000 people.

The Tambora volcano was one of the largest known ash-producing eruptions in the last 10,000 years. An estimated 36 cubic miles of ash and pumice, equal to about 12 cubic miles dense rock equivalent (DRE) of magma, was ejected. Ash fallout was noted over an estimated area of 1/2 million square miles. Darkness was observed to last for up to two days at distances of more than 350 miles from the volcano. Interestingly, fallout from the Tambora eruption has also been detected in Antarctic ice cores.

GLOBAL EFFECTS

The eruption rate and the area of ash dispersal both suggest that the eruption column may have reached as much as 30 miles into the stratosphere. The volcanic cloud traveled around the world, and, within three months, its optical effects were observed at distant locations in Europe. For example, around the end of June, and later in September, several observers near London reported prolonged and brilliantly colored sunsets and twilights. Atmospheric effects began to appear in the skies above the northeastern United States in the spring of 1816. The dominant effect was a persistent dry fog or dim sun. The haze was clearly located above the troposphere, because neither surface winds nor rain dispersed it.

In Maryland, people knew something strange was happening when the snows of late spring were brown, blue, and even red instead of white. Brown snow also fell in Hungary and in southern Italy, where any snow is considered unusual; red and yellow snows that season caused considerable alarm.

In New England, 1816 was called "the Year Without a Summer" and also, with typical Yankee wryness, "Eighteen Hundred and Froze to Death." As the spring wore into summer, there were successive cold waves and frosts every month. Thanks to the presidents of Yale University during this period we know just how cold the summer was in New Haven, Connecticut. It was the habit of the Yale presidents, a succession of clergymen and scholars, to rise each morning at 4:30 a.m. to read and record the temperature. In June 1816, the average temperature was found to be 7°F below the average June temperature for the preceding years and the years that followed. A study completed in the 1970s demonstrated that the summer of 1816 was the coldest summer in New Haven, Connecticut, for the entire period from 1780 to 1968. While similar records of temperature had been maintained at Harvard prior to the summer of 1816, an unfortunate gap in the record keeping just at this critical time occurred when the keeper of the records, one Samuel Williams, fled to Vermont to avoid prosecution for embezzling funds from a Harvard trust.

Further north, the records of the Hudson Bay show that the summers of 1816 and 1817 were the coldest of any in the modern record. Tree ring data from northern and western Quebec support these observations.

Crop failures were rife in New England the summer of 1816. Only 1/4 of the principal staple crop, Indian corn, ripened sufficiently to be used for meal, and hay and wheat crops were in similar straits. Fortunately, the United States was an exporter of agricultural products at that time, so there was typically a surplus beyond what was needed for local consumption.

What was a major inconvenience in America was a disaster in many parts of Western Europe, just starting to emerge from the chaos of the Napoleonic

Wars and far more dependent on immediate local agricultural production. Crop failures were universal, resulting in food shortages and famine in many locations. In some districts of Germany and Switzerland, officials sealed off their borders to prevent the export of grain. By the end of 1816 in Ireland, the deficient harvest of the summer had been completely consumed, and small landowners were forced to abandon their homes by the spring of 1817 and beg for a living. Throughout Europe food riots broke out, armed groups raided farms, and bakeries and grain markets were looted. Some landowners fortified their estates against roving bands of the destitute.

It has been suggested that the conditions of famine promoted the activities of the typhus-bearing louse responsible for the European epidemic between 1816 and 1819. In Ireland alone during this period, an estimated 1.5 million people were afflicted with typhus, leading to 65,000 deaths.

MODERN-DAY EFFECTS FROM TAMBORA

In addition to the climatic affects of the Tambora eruption, there were some interesting side effects; one of these involved the English novelist Mary Shelley. In the summer of 1814, the poet Percy Bysshe Shelley eloped to France with then Mary Godwin. Their official, sanctioned marriage, however, would have to await the death of Percy's wife, Harriet, in December 1816. During May of 1816, Percy and Mary traveled to Lake Geneva to summer with their friend, poet Lord Bryon. Forced to stay indoors by the unseasonably cold and wet weather caused by Tambora, the group decided to have a ghost-writing contest. Mary began writing a horror story that would become the novel *Frankenstein*, which she finished in the spring of 1817. While Percy, Lord Bryon, and other guests wrote stories for the contest, the success of *Frankenstein* would endure to this day, long after the other writings produced that summer had faded.

A second result of the Tambora eruption was the invention of the bicycle. In the German state of Baden, a series of crop failures that began in 1812 had resulted in an exorbitant price for oats. This prompted an inventor, Karl Drais, to find a replacement for the horse. Horses hauled almost everything that moved, and must be fed whether they were working or not. Drais' first design, called a velocipede, consisted of four wheels. His second design was a two-wheeler based on the key principle of modern bicycle design: balance. Today we think of balancing on two wheels as relatively easy to learn. This was not the case for society in the early 1800s, where people at that time normally only took their feet off the ground when riding horses or sitting in a carriage. Drais recognized this and decided to play it safe in his design. Instead of propelling the machine, called a draisine, with a crank driven by the feet, riders simply scooted by pushing with their feet. While the evidence linking Drais's invention to the eruption of Tambora is circumstantial, it is still considered persuasive. Contemporary newspapers certainly hinted at a link. In 1817, the *Dresdner Anzeiger* reported that as the draisine replaced horses, the hope was that the price of oats would fall.

KRAKATOA

While the Tambora eruption was one of the three or four largest explosive volcanoes in recorded history, 68 years later a smaller cousin would erupt along the Indonesian archipelago and end up contributing more to our understanding of the impact of volcanic activity on the global atmosphere than any preceding event. The eruption was the famed Krakatoa, which we repeat is still west, not east, of Java.

There are many in-depth accounts of the aftermath of Krakatoa. One of the most descriptive and eminently readable is that by Simon Winchester, *Krakatoa: The Day The World Exploded August 27, 1883*. Rather than repeat accounts of the horrors of this event, we will focus on the most important effects as they relate to the theme of this chapter.

While estimates of the size of Krakatoa place it at about 1/5 the size of Tambora, the eruptions in August 1883 still produced disastrous results in the region. The official death toll recorded by the Dutch authorities was 36,417, the majority of deaths a result of a series of five tsunamis caused by five massive pyroclastic flows entering the waters surrounding the volcano on the morning of August 27. The largest tsunamis exceeded 100 feet in height.

While the death toll and the climatic effects reported for Krakatoa were less than Tambora, the real significance of Krakatoa lies in the scientific realm. For the first time, the global atmospheric effects of a large explosive eruption would be studied and documented in a modern scientific fashion. The eruption occurred in the middle of a busy shipping lane, so there were many well-documented eyewitness accounts of the eruptions. And additionally, with the establishment of global communications, accounts of the eruptions spread throughout Europe and North America within days of the major eruptions on August 27. The fact that global communications were considerably more limited in 1815 probably accounts for the lack of notoriety of Tambora, compared to the fame, or infamy, heaped on Krakatoa. This is really unfortunate for the Hollywood movie director, Bernard Kowalski, who directed the mistitled epic *Krakatoa, East of Java*, because, if Tambora had taken its rightful place in the halls of infamy, a film *Tambora, East of Java* would at least have been geographically correct if nothing else.

Two historically significant scientific investigations were initiated following the August 1883 eruptions of Krakatoa. The first investigation was commissioned by the Dutch Indian government in October 1883, headed by a mining engineer by the name of R.D.M. Verbeek. Verbeek had mapped the island of Krakatoa three years prior to the eruptions. He published a 546-page monograph in 1885 describing the geology and the effects of the eruptions on the immediate vicinity.

By the beginning of 1884, it was obvious that the explosive eruption of Krakatoa had created worldwide atmospheric effects. In response to this, the Royal Society of London appointed a committee to examine the effects. Four years later, the Royal Society released its report. The results presented in the report were based on data collected from more than 50 observatories around the world, and identified a number of atmospheric phenomena, ranging from reports of loud noises at extreme distances to a variety of optical phenomena.

The final explosive eruption that occurred around 10:02 a.m. on the morning of August 27 was one of the loudest sounds in recorded history, the eruption being heard up to 3,000 miles away—more than 400 miles farther than the distance between San Francisco, California, and New York, or approximately twice the distance between Paris, France and Moscow, Russia. On the island of Rodriguez located in the Indian Ocean, 3,000 miles from Krakatoa, there was a report of the distant roar of "heavy guns" heard coming from the east several times during the night.

Amazingly, there is also a report of sounds from the final stages of the eruption reaching a level of 180 dBSPL (sound pressure level), 100 miles from the volcano. The exact source of this report is unclear, but anyone who survived the tsunamis and was around to experience this sound level would have been in excruciating pain, most likely followed by permanent hearing damage and ruptured eardrums. A typical hearing chart indicates that the threshold for pain starts at about 130 dBSPL and that sound measuring 180 dBSPL corresponds to the sound of 1 ton of Trinitrotoluene (TNT) exploding at a distance of 250 feet, which, admittedly, is still hard to imagine, but at least conjures up something we can begin to comprehend.

For three years after the final explosive eruption of Krakatoa on the morning of August 27, 1883, many parts of the world were treated to a variety of atmospheric effects linked to the eruption. In many places, the days were filled with a blue or green haze, with spectacular red glows arising just after sunset and just before dawn. So intense was the glow that 16 months after the eruption people in Poughkeepsie, New York, on the morning of November 27, 1884, thought that the red glow was from a fire, and called the fire department. *Tales of the Earth: Paroxysms and Perturbations of the Blue Planet*, by Charles Officer and Jake Page, has an excellent discussion of the cause of the after sunset and before dawn refractive behavior of sunlight as seen through the prism of the volcanic dust and aerosols of Krakatoa. Other atmospheric phenomena associated with the eruption included noticeable dimming and blur ring of celestial objects and unusually dark lunar eclipses.

Accurate inferences about high-altitude atmospheric winds and about the thickness of the supposed "dust cloud" were soon drawn by following the progressive changes of the striking visual phenomena. Today, we recognize that the observers were making prediscovery observations of the stratosphere under the influence of a large volcanic eruption. In essence, they were further defining the *what* so that later researchers could better address the *why*.

More than 80 years would pass before the next step forward in our understanding of the impacts of explosive volcanic activity on climate and the environment would occur. The Agung stratovolcano, located on the Indonesian island of Bali would provide that step.

Agung is Bali's highest and most sacred mountain. The 2-mile-high summit of this volcano contains a steep-walled, 1,640 feet-wide, 656-feet-deep crater. Agung had been dormant for nearly 150 years when on March 17, 1963, it erupted catastrophically. The explosions continued for about 7 hours, and by the time they ceased, pyroclastic flows, ejected rock fragments, ashfalls, and lahars were responsible for the deaths of nearly 1,150 people and the injury of 300 more. In addition to the human death toll, the volcanic eruption destroyed hundreds of homes, and buried nearly all the crops under ash and mud.

The explosive eruption of the Agung volcano provided the evidence that tipped the scale in favor of sulfur from volcanic gas, as opposed to ash or dust, as the key to understanding how and why volcanic eruptions impact climate. Volcanologists and climatologists had, for a long time, suspected that gases and ash propelled high in the atmosphere by explosive volcanic eruptions could partially block the sun's rays and cause a cooling at the Earth's surface. Direct measurements made at the surface and well into the stratosphere in the wake of Agung showed convincingly that material from Agung had reached the stratosphere, and that at the same time the troposphere had cooled. The observations indicated that temperatures in the stratosphere fell by about 10.8°F and that the 1963–1966 average temperature in the Northern Hemisphere fell by about 0.7°F. Sulfur from Agung circled the globe many times and eventually showed up as conductivity spikes associated with sulfuric acid in ice cores drilled in Greenland.

In 1978, climatologists at the Goddard Institute for Space Studies, in a paper published in *Science*, stated that the Agung eruption provided the best-documented global data on the atmospheric effects of gases from a volcanic eruption. In the concluding paragraph of this paper, the authors made statements that proved to be prophetic and are repeated here:

Finally, we emphasize the enormous potential scientific value of the natural climatic experiment provided by a large volcanic eruption. One very useful strategy for investigating the global climatic system would be to make detailed observations and analyses of the next large volcanic explosion. The observations should include global monitoring of the spread, growth, and decay of the aerosols, sampling of the aerosol properties in situ, and accurate global monitoring of the climatic response. These data would permit testing of global climate models and aid in-depth analyses of radiative, chemical, and dynamical processes, which are essential for obtaining an improved understanding of the physical mechanisms and interaction involved.

EL CHICHON

They wouldn't have to wait long! In late March and early April of 1982, the volcano El Chichon, located in the southeastern region of Mexico, came to life after more than a century with an explosive eruption measuring 5 on the VEI scale. The El Chichon eruption sent a huge cloud rich in sulfur dioxide to an altitude of about 16 miles into the stratosphere. Observations from instruments on the ground, in airplanes, and by satellite showed that this eruption had a large effect on the stratosphere, and that the spread of the eruption cloud could be accurately tracked.

El Chichon provided the test case for which volcanologists and atmospheric scientists had been waiting. While the volcanic ash ejected into the atmosphere from this volcano was small, the amount of sulfur dioxide was considerable, estimated at about 7 Mt. The reaction of the sulfur dioxide with water vapor and other gases in the stratosphere produced more than 10 Mt of sulfuric acid aerosols on a timescale of weeks. According to modeling results, this amount of aerosols was enough to lower the surface temperature in the Northern Hemisphere several tenths of a degree Fahrenheit and temperature measurements for the year 1982 showed such a cooling corroborating the theoretical modeling results.

While El Chichon and the earlier eruption in 1980 at Mount St. Helens were both assigned VEIs of 5, the similarity stops there. An estimate of the amount of sulfate aerosols formed in the stratosphere from Mount St. Helens was between one and two orders of magnitude less than that for El Chichon; 0.3 Mt versus 20 Mt. This difference in stratospheric aerosols was, in part, a result of the fact that Mount St. Helens was predominantly a lateral blast, whereas El Chichon was vertical. Thus, we see that similarity of VEI index does not necessarily imply similar climatic effects.

MOUNT PINATUBO

Nine years later, one of the most important eruptions in terms of advancing our understanding of the interaction of an explosive volcanic eruption and the atmosphere occurred in the Far East. During the month of June 1991, Mount Pinatubo, located on the island of Luzon in the Philippines, underwent a series of explosive eruptions culminating on June 15 in the second largest eruption of the 20th century, measuring 6 on the VEI scale. The climactic eruption on June 15 was the culmination of a 9-hour series of plinian and pyroclastic-flow-producing eruptions resulting in the production of approximately 1 cubic mile of ejected material.

The eruption of Mount Pinatubo resulted in an immediate fallout of material near the volcano itself and, more importantly, injected approximately 20 million tons (Mt) of sulfur dioxide into the stratosphere to an altitude of about 22 miles. The photograph was taken from Clark Air Force base on June 12, 1991, about 16 miles from the volcano. It is one of the clearest pictures of a

Fig. 3-3. A huge cloud of volcanic ash and gas rises above Mount Pinatubo in the Philippines on June 12, 1991. Three days later, the volcano exploded in the second-largest volcanic eruption on Earth in this century. Timely forecasts of this eruption by scientists from the Philippine Institute of Volcanology and Seismology and the U.S. Geological Survey enabled people living near the volcano to evacuate to safer distances, saving at least 5,000 lives. U.S. Geological Survey photograph taken by Dave Harlow.

plume, or cloud, of volcanic ash and gases jetting into the atmosphere. While the larger climatic eruption occurred on June 15, the vantage point for this photo, Clark Air Force base, had been evacuated by that time for safety reasons. It is also highly probable that Mount Pinatubo was obscured by eruption clouds by the time of the climatic eruption on June 15, and a photo as clear would not have been possible.

"For the past 15 years, atmospheric scientists have taken advantage of the myriad of ground, air, and space-based observations and measurements of a wide spectrum of effects caused by this volcano to test and refine models of the Earth's atmosphere and, more importantly, the climate system. Between 2001 and 2006, 118 scientific presentations addressing different aspects of the Pinatubo volcano were delivered before hundreds of scientists at the fall annual meeting of the American Geophysical Union. Several of these presentations dealt with the latest findings of the imapct of Mt. Pinatubo on global warming and ozone depletion, two areas of prime importance to humans. As Jim Hansen, world famous climatologist at the NASA Goddard Institute for Space Studies, remarked, "It is the best volcanic test of the past century, because, at least in our calculations, it is the largest volcanic forcing since Krakatoa in 1883, and the aerosol properties were observed much better than those of any other large volcano."

The climatic Pinatubo cloud was the largest sulfur dioxide cloud ever observed in the stratosphere since the beginning of satellite observations in 1978. More than three months after the eruption, pictures taken at a height of 16 miles by the Upper Atmospheric Research Satellite showed that the cloud still stretched around the globe.

The 20 Mt of sulfur dioxide injected into the stratosphere immediately began to convert into sulfuric acid (H_2SO_4) aerosols, forming the largest perturbation to the stratospheric aerosol layer since the cloud of Krakatoa in 1883. The aerosol consisted of an estimated 25 to 30 Mt cloud of sulfuric acid droplets. The droplets are similar to little cloud droplets that move around and gradually fall out of the stratosphere due to the influence of gravity. The stratospheric aerosols spread rapidly and encircled the Earth in about three weeks, and attained global coverage about one year after the eruption. This large aerosol cloud impacted the climate on several continents.

The winter of 1991–1992 turned out to be very warm over most of North America, Europe, and Siberia, but cold over Greenland and the eastern

Mediterranean. It snowed in Jerusalem that winter, and corals at the bottom of the Red Sea died because of the cold surface waters. During the summer of 1992, continental regions of North America and Eurasia were unusually cold, with temperatures up to 4°F cooler than normal. Global warming halted for several years as the planet cooled temporarily. Also, during this time, cold temperatures in Greenland led to a significant reduction in the summer surface melt in 1992 and 1993 across the southern Greenland Ice Sheet, a trend which, unfortunately, has since reversed.

In November 2005, it was announced in a USGS Volcano Watch weekly publication that recent research suggests that volcanic eruptions possibly play a more important role in climate modification than previously suspected. While the article states that volcanic eruptions might be keeping sea level rise (caused by global warming), in check, it goes on to explain that "in check" means temporarily moderated. For instance, it has been estimated that the eruptions of Mt. Pinatubo in 1991, Mexico's El Chichon in 1982, and Indonesia's Agung in 1963 reduced sea level rise by about 0.28 inches—only a fraction of the 7.1 inch increase in sea level observed in the 20th century. Thus, the article concluded, volcanic eruptions can only temporarily delay the effects of sea level rise caused by global warming—they cannot halt the process or change the long-term rate of change.

GLOBAL WARMING

The Mount Pinatubo eruption also plays an important role in a stratospheric albedo enhancement scheme proposed by Professor Paul Crutzen of the Department of Atmospheric Chemistry, Max-Planck Institute for Chemistry, Mainz, Germany, and the Scripps Institution of Oceanography, University of California, San Diego, California. The scheme is described in a thought-provoking article in the August 2006 *Journal of Climatic Change*. Professor Crutzen was a co-recipient of the 1995 Nobel Prize in Chemistry for his work in atmospheric chemistry, particularly concerning the formation and decomposition of ozone. The objective of the albedo enhancement or modification scheme is to mitigate global warming by artificially introducing sulfur into the Earth's stratosphere. It's safe to say that the technique for getting the sulfur into the stratosphere is not finalized. One possibility mentioned in the climate change article would be to carry the sulfur into the stratosphere on balloons and use artillery guns to release it (wow). The sulfur would promote the formation of sulfate aerosols which, as we know from observations of the explosive volcanic eruptions of the 20th century, would reflect incoming solar radiation back into space and result in a cooling of the Earth's surface. Many of the important parameters needed for such a scheme are based on observations for the Mount Pinatubo eruption, including an estimate of the amount of sulfur dioxide injected into the stratosphere, the observed average global temperature drop one year after the injection, and an estimate of the radiative cooling, given the amount of sulfur that remained in the stratosphere as sulfate several months after the eruption.

In his article in the *Journal of Climate Change*, Professor Crutzen cautions that his stratospheric albedo modification scheme should only be considered as a last resort. The best solution to the global warming problem would be a significant reduction in the carbon dioxide emissions. However, "given the grossly disappointing international political response to the required greenhouse gas emissions…research on the feasibility and environmental consequences of climate engineering of the kind presented in this paper, which might need to be deployed in the future, should not be tabooed." He adds: "the possibility of the albedo enhancement scheme should not be used to justify inadequate climate policies but merely to create a possibility to combat potentially drastic climate heating."

The Mount Pinatubo volcano resulted in significant destruction of the stratospheric ozone layer in the years following the eruption. Ground and space-based measurements indicated decreases as large as 20 percent at altitudes of 1 to 10 miles in the tropical stratosphere 3 to 6 months after the climatic eruption on June 15. By the time of maximum aerosol development, a reduction of up to 20 percent was measured over Colorado and Hawaii, and mid-latitude ozone levels reached their lowest levels ever recorded during 1992 to 1993. The globally averaged ozone reduction was 2 to 3 percent, lower than in any previous year for which measurements were available, with the largest decreases in the regions from 10°S to 20°S latitude and 10°N to 60°N latitude.

While chlorine had been identified as the ozone-destroying element in the Earth's stratosphere, at the time of the Pinatubo eruption there was some debate over its origin. Some scientists argued for the injection of chlorine into the stratosphere by volcanoes as the primary source. This led to speculation about what might happen if a volcano the size of Tambora in 1815, with an estimated aerosol mass 5 to 10 times that of Pinatubo, were to occur. If the response of the stratospheric ozone layer were proportional, there could be drastic increases in the amount of harmful ultraviolet radiation (UV-B) reaching the Earth's surface. In that event, the effect on plant life and potential increase in skin cancer and cataracts in humans could be catastrophic.

Multiyear data from NASA's Upper Atmospheric Research Satellite (UARS) ended that debate. The UARS instruments found chlorofluorocarbons (CFCs) in the stratosphere and traced the worldwide buildup of stratospheric gases corresponding to the breakdown of CFCs, showing that it was not volcanoes that were responsible for these gases. While volcanic eruptions emit hydrogen chloride, a potential source of chlorine, they also spew out huge amounts of water vapor. Because hydrogen chloride readily dissolves in water, most of the chlorine emitted from a volcanic eruption never reaches the ozone layer in the stratosphere. It is, instead, washed out of the atmosphere in rain. Observations of the El Chichon and Pinatubo eruptions demonstrated that these eruptions produced very small increases in stratospheric chlorine.

CFCs

Chlorofluorocarbons, or CFCs, once used in a variety of products from air conditioners to refrigerators to underarm deodorant sprays, were considered safe because they formed an inert gas that didn't react chemically with anything. This indestructibility of CFCs allows them to remain airborne until they eventually drift into the stratosphere. Sulfate aerosols formed in the stratosphere following an explosive volcanic eruption interact with the CFCs, leading to the enhanced destruction of ozone.

Ozone depletion by volcanic aerosols is a recent phenomenon. It is caused by elevated chlorine concentrations in the stratosphere, which only appeared in the last few decades because of manmade emissions, and will hopefully disappear in a few decades as emissions are increasingly regulated. No obvious effects associated with depletion of the ozone layer, such as reports of increased skin cancers and cataracts, were reported after the massive Tambora eruption in 1815 or the 1883 eruption of Krakatoa. This is consistent with the fact that there were no CFCs in the stratosphere at the times of those large eruptions, as none of the conveniences of modern living that involve the use of CFCs were invented or in widespread use. An additional possibility for the lack of reports may be related to the scarcity of medical expertise and facilities.

CFCs produced between the 1930s and 1987, when 68 nations signed the Montreal Protocol banning CFC gases, are expected to be around for some time. One estimate indicates that approximately 50 years from now, there will still be as many CFCs as there were in the late 1980s. While the possible occurrence of a volcanic eruption on the scale of Mount Pinatubo (VEI 6), or larger, will continue to pose a major threat to the ozone layer until the mid-21st century or longer, recent findings obtained serendipitously could mean that significant destruction of the ozone layer is even more possible than thought just a few years ago.

In 2000, a DC-8 loaded with operational instruments for measuring particle concentrations in the atmosphere was flying over Iceland when it accidentally flew through a plume thrown up by a relatively small eruption of the Icelandic volcano named Hekla. Scientists from the United Kingdom reported that volcanic gases from Hekla that penetrated the stratosphere were, in addition to sulfates, forming ice and nitric acid particles. This is a critical finding, as these latter particles "switch on" volcanic chlorine gases, accelerating reactions that lead to ozone destruction and the formation of mini-ozone holes.

SUPERVOLCANO EFFECTS ON THE OZONE LAYER

As discussed in Chapters 5 and 9, one of the potentially most deadly effects of a massive explosive injection of volcanic gases into the stratosphere by a supereruption, in contrast to the much smaller eruptions discussed so far, is the possibility of a significant depletion of the ozone layer. Ilya Bindeman and others state, in the abstract of a paper given at the December 2006 fall annual meeting of the American Geophysical Union (AGU), that results of their study of the oxygen isotope geochemistry of ash particles from supereruptions like Yellowstone in Wyoming and Long Valley in California indicate that these massive eruptions may result in drying out the stratosphere from water, and cause the temporal depletion of the ozone layer. They conclude that the magnitude of the depletion may be many times that of the measured 3 to 8 percent depletion following the 1991 Mount Pinatubo explosive eruption depending on how ozone depletion scales with the amount of sulfur dioxide released into the stratosphere.

VOLCANOES IN RECORDED HISTORY

So far the focus has been on the volcanic impact on climate. We now turn to the impacts of several of the most violent volcanic eruptions in recorded history both on the environment and civilization at the time of the eruptions. Historical accounts from many locations in different parts of the world point to the occurrence of massive volcanic eruptions in the 6th and 13th centuries. These events, in addition to severe climatic effects, had major, and in one case, life-changing, demographic repercussions over much of Europe, the Middle East, China, and areas as remote as Siberia.

EL CHICHON

The first of these volcanic events occurred in A.D. 1258 and is considered to have been one of the largest explosive eruptions in the past 10,000 years. Estimates of the size of this eruption place it at approximately twice the size of the 1815 Tambora eruption in terms of the amount of ejected material. While the identity of the specific volcano responsible for this massive eruption remains a mystery, as we have seen from more recent observations of events in the 1800s and 1900s, a tropical location is considered likely, given the worldwide presence of ash and the simultaneous occurrence of prominent sulfuric acid signals picked up by recording devices in ice cores from Greenland and Antarctica. Sulfate levels in the ice cores were two to three times higher than levels for Tambora, which as we saw earlier, produced the "year without a summer." A candidate site for the A.D. 1258 eruption is El Chichon in southern Mexico.

Aerosols in the stratosphere from the 1258 eruption produced a dry fog thought to be similar to the "dry fog" reported by Benjamin Franklin after the 1783 Laki eruption in Iceland. The fog dimmed the sun's light, chilled the atmosphere, and, according to a study completed in 2006, likely triggered a moderate-to-strong El Nino event in the midst of prevailing La Nina-like conditions. Historical evidence from the time of the eruption points to massive rainfall anomalies, with adverse effects on agriculture and ensuing widespread famine. A fatal disease pandemic originating from a focus in Asia or Africa spread throughout the Mediterranean area within one to five years after the eruption. The contagion responsible for the mass mortality is thought to have been plague.

A connection between famine and plague in the 13th century has been explained as the result of the strong dependence of world economies on short-term agricultural output and, as a result, the vicissitudes of weather. Thus, poor crop yields following on the heals of volcanically produced dry fogs would result in severe famine which, in turn, would lead to disease in both animals and humans due to lowered body resistance. Plague is thought to have been an indirect effect of famine. Hungry, plague-bearing rats driven from the fields to seek out grain stockpiles would come into intimate contact with humans, thereby spreading the disease.

In recent years, there has been speculation that the occurrence of a number of moderate-to-large (that is, VEIs 6 to 7) explosive eruptions spaced over years or decades could result in a long-term impact on global climate. One of the mechanisms for such an impact could be colder temperatures and increased precipitation, causing a substantial increase in ice and snow cover and a resultant increase in the Earth's albedo. This kind of a chain reaction would act as a feedback mechanism enhancing and prolonging the climatic effects. In the austral summer of 2000–2001, two ice cores were drilled to a depth of 403 feet at the Amundsen-Scott South Pole Station, Antarctica.

Analysis of these new cores, combined with results of analyses of older Antarctica cores, showed that just such a scenario of moderate to large explosive eruptions, five in all, occurred within a 50-year time span in the 13th century, with the A.D. 1258 event being the largest. Based on the amplitude of the sulfuric acid signals in the ice cores, it was determined that the amount of material ejected during the 13th century was a factor of 3 to 20 times larger than the amount of material ejected in any century in the previous 1,000 years. Some scientists also suggested that the frequent injection of large amounts of volcanic aerosols into the atmosphere during the 13th century, combined with a gradual diminishing of sunspot activity, may have contributed to the climatic transition from the Medieval Warm Period to the Little Ice Age that started either late in the 13th or early in the 14th century.

PROTO KRAKATOA

The densest and most persistent stratospheric dust cloud, or "dry fog," in recorded history was observed over Europe and the Middle East during A.D. 536 and A.D. 537. The importance of this cloud lies in the fact that its mass and its climatic consequences appear to exceed those of any other volcanic cloud observed during the past 3,000 years. While some researchers differ in their estimates of the exact time of occurrence and location of the responsible volcano, a remarkable piece of investigative reporting and modeling work at one of the country's leading scientific laboratories rather convincingly points to A.D. 535 and the Indonesian archipelago for the timing and location, respectively, of the event.

The investigative reporting comprises the book, *Catastrophe: An Investigation into the Origins of the Modern World* by David Keys, and the modeling work conducted by Ken Wohletz, a volcanologist at the Los Alamos National Laboratory (LANL) is summarized in a report titled, "Were the Dark Ages Triggered by Volcano-Related Climate Changes in the 6th Century? (If So, Was Krakatau Volcano the Culprit?)." Note that the LANL report is available online.

It started with the trees: tree rings that is. Dendrochronologists, experts in dating events and variations in environment based on the width of growth rings in trees, were amassing data that pointed to a time in mid-6th century A.D. when tree rings from trees around the world exhibited similar, but strikingly anomalous, behavior compared to the past roughly 2,000 years or so. The tree ring data came from locations as far apart as Chile, the Sierra Nevada mountains in California, Finland, Ireland, and Siberia, and all indicated the same thing: a drastic drop in summer growth starting around A.D. 535 and continuing into the A.D. 540s. The evidence was in: The implication was of extremely short or no summer growth and long stretches of extreme cold in the mid-6th century.

During the next two to three years, historical documents from the far corners of the 6th-century world were obtained and studied for evidence of a climatic catastrophe and the impacts on civilization at that time. The findings of this multi-year effort are detailed in David Keys's book. His genius is in the integration of "hitherto unrecognized connections between the wasteland that overspread the British countryside and the fall of the great pyramid-building Teotihuacan civilization in Mexico, between a little-known Jewish empire in Eastern Europe and the rise of the Japanese nation-state, between storms in France and pestilence in Ireland." The next step was to identify the cause of this cataclysm.

The work of several scientists contributed to the identification of the causative event: a massive explosive volcanic eruption in the Indonesian archipelago. In particular, a comprehensive analysis of a caldera formation and atmospheric interaction was carried out for a hypothetical massive Proto Krakatoa event by the previously-mentioned LANL volcanologist, Ken Wohletz. By combining information from tree rings, ice cores, eyewitness accounts of the dimming of the sun, and bathymetric estimates of a caldera in the sea floor of the Sunda Straits, supercomputer simulations of an explosive eruption were carried out. From the assumed size of the caldera (that is, about 31 miles in diameter), the amount of magma erupted was estimated at 48 cubic miles. As discussed in Chapter 4, this volume of material places the Proto Krakatoa event just below the level of a supervolcano, but still a massive event.

Extremely important aspects of the Proto Krakatoa event were: first, the formation of the 30 miles diameter caldera, with collapse below sea level, could have formed the Sunda Straits separating Java from Sumatra as suggested by ancient Javanese historical writings; and second, the caldera collapse would have allowed inpouring of tremendous amounts of sea water from the Indian Ocean and the Java Sea. Computer simulations of the sea water magma interaction demonstrate that a plume from 15 to greater than 30 miles high, carrying from 12 to 24 cubic miles of vaporized seawater into the atmosphere, could have been produced. The plume would loft large quantities of water vapor into the stratosphere, forming a globally distributed layer from 65 to 490 feet thick of ice clouds with fine hydrovolcanic ash. The impact of the plume on the Earth's albedo and stratospheric ozone with the release of chlorine from the sea water are unknown, but could result in global climate destabilization lasting years or several decades. Thus, a Proto Krakatoa event could account for the far-reaching environmental and ecological effects that are described in the historical accounts and preserved in the tree rings and ice cores.

SANTORINI ARCHIPELAGO

Around 1630 B.C., a plinian-style explosive eruption occurred on the Greek island of Thera in the Santorini archipelago. This volcanic eruption has been linked to the downfall of the highly sophisticated Minoan civilization, the disappearance of the fabled Atlantis, and the exodus of the Jewish people out of Egypt.

The Santorini archipelago is a small group of volcanic islands in the Aegean Sea, located about 125 miles southeast from the mainland of modern-day Greece. It is also known by the name of the largest island in the archipelago, Thera. It is the most active volcanic system in the Aegean arc with the larger of the volcanically active Kameni Islands, Nea Kameni, the site of lavas less than 50 years old. The name *Santorini* was given to the island group in the 13th century, and is a reference to Saint Irene (or Irini), patron saint of sailors. Before that time, the islands were called Kalliste ("the most beautiful one"), Strongyle ("the circular one"), or Thera.

As seen in the following satellite image, a giant central embayment, more or less rectangular in shape and measuring about 8 miles by 4 miles, is

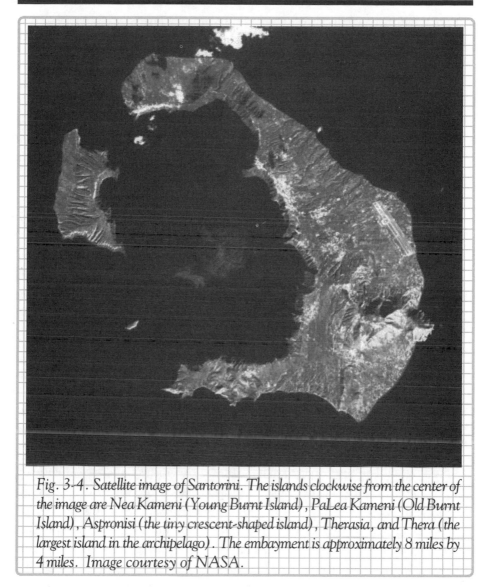

Fig. 3-4. Satellite image of Santorini. The islands clockwise from the center of the image are Nea Kameni (Young Burnt Island), PaLea Kameni (Old Burnt Island), Aspronisi (the tiny crescent-shaped island), Therasia, and Thera (the largest island in the archipelago). The embayment is approximately 8 miles by 4 miles. Image courtesy of NASA.

surrounded by 1,000-foot-high sheer cliffs on three sides. On the fourth side, the lagoon is separated from the Mediterranean by the smaller ring island, Therasia. With water depths of about 1,300 feet in the center of the embayment, it is an ideal safe harbor for even the biggest ships. The island's ports are all located inside the lagoon to take advantage of the deep water, shelter from storms, and high seas in the Mediterranean. The towns of Santorini cling to the top of the cliffs looking down on the embayment. The spectacular beauty of the rugged volcanic islands, the deep-water embayment, and the famed nightlife make these islands some of Europe's top tourist attractions.

The embayment is known to be the result of major caldera-forming events, the most recent being Thera, which scoured out the northern half of the embayment. Pre-eruption Thera was a large island dominated by a water-filled embayment to the south and a central highland to the north. During the Thera eruption, the central highland collapsed to form the full extent of the present-day caldera and the three islands of the Santorini archipelago.

The results of archaeological research reported in 2006 by a team of international scientists indicated that the Thera event was even more massive than previously thought. It is now estimated that the eruption expelled 15 cubic miles of magma and rock into Earth's atmosphere, compared to estimates in 1991 of only 9 cubic miles. The volume of magma ejected by the Thera eruption is estimated to have been between four and six times what was blown out by Krakatoa in 1883. The VEI of the Thera eruption is estimated at 6.9; in the past four millennia only seven eruptions are known to have had VEI values that were equal or larger than Thera. Age data for the extensive pyroclastic formations on the island of Thera suggest a recurrence interval for explosive eruptions of the size of Thera of about 20,000 years, with smaller explosive eruptions every 5,000 years. On a global basis, explosive eruptions the size of Thera occur approximately every 300 years.

The eruption sequence for the Thera event has important implications for archeological findings on the island. Studies throughout the past several decades have identified four major eruption phases and one minor previous tephra fall. The second and third major eruption phases involved the incorporation of significant quantities of sea water. The fourth, and last, major phase consisted of pyroclastic flows and lahars, and was characterized by extensive flooding of the newly formed caldera that currently occupies the northern half of the embayment shown on the map and satellite image. The archeological site has been described as "the Pompeii of the Late Bronze Age." The immensity of the eruption is evident from widespread ash deposits in deep-sea sediments of the Aegean, Mediterranean, and Black Seas, as well as in lake sediments in Turkey. Ash deposits have also been found at archeological sites as far east as Cyprus and the Nile delta, some 500 miles distant.

On the southern coast of Thera near Akrotiri, one of the most important prehistoric settlements of the Aegean is being exhumed from the tephra deposited by the Thera eruption. One of the most interesting findings coming from the archeological dig at Akrotiri is that no human or animal remains have been found beneath the tephra in the otherwise well-preserved settlement. This is in striking contrast to the digs at Pompeii and Herculaneum in Italy, where much of the population was interred by volcanic deposits from the A.D. 79 eruption of Vesuvius. In addition, the inhabitants of Akrotiri had removed most of their belongings, clearly indicating the town was abandoned before the start of the first four major eruptive phases of Thera. Archeological evidence uncovered at the site suggests that the reasonably intelligent and well-to-do people of Akrotiri heeded the advance warning sign of the minor precursor activity and got out of town before the onset of the first major eruptive phase.

It has been suggested that the Thera eruption triggered the downfall of the Minoan civilization on the island of Crete 70 miles to the south. In a classic plinian eruption marked by columns of smoke and ash extending high into the stratosphere, the Thera eruption created a plume 18 to 21 miles in height, and magma coming into contact with the shallow marine embayment would have caused a violent phreatic eruption (a steam eruption caused by the interaction of hot magma with groundwater). The eruption also generated a 115- to 500- foot-high tsunami (estimates vary) that devastated the north coast of Crete approximately 70 miles away. The impact of the tsunami pummeled coastal towns where building walls were knocked out of alignment. The tsunami would also certainly have eliminated every timber of the Minoan fleet along Crete's northern shore. Although many inhabitants would have died, many would have survived, but only to face the longer-term effects of the volcano, which would have eventually weakened them to the point where they were vulnerable to attack from outside forces. With a gradually weakening sea-faring economy, the Minoan culture on the island of Crete gave way to the mainland Greek Mycenaean culture.

Archeology has shown that the Minoan culture was probably one of the most sophisticated and advanced cultures, not only of its time, but for centuries after its disappearance. The Minoans dominated the eastern Mediterranean with a powerful navy, and may have extracted tribute from other surrounding nations. Excavations in Crete, the seat of the Minoan Empire, have uncovered evidence of unparallel architecture, art, a code of laws giving women equal status to men, and a highly developed agriculture based on an extensive irrigation system second to none. Then, seemingly in the geological blink of an eye, the Minoan civilization disappeared.

Many people of varied backgrounds believe (and believe is the operative word) that Thera once was the fabled Atlantis. As is well known, Atlantis was a prosperous land that, according to the ancient Greek philosopher Plato, disappeared without a trace, sunk into the sea by anger of the gods. Volcanologist Heraldry Sigurdsson of the University of Rhode Island, lead author on an article published in the August 22, 2006 issue of *Eos* that describes the latest results from a marine geology survey of the ancient Thera volcano, is convinced that the explosive eruption played a big role in the mythology of the ancient Greeks. He is quoted in an article in the August 27, 2006 issue of *USA Today* as follows: "In my mind, personally, the mythology born out of this largest eruption must be responsible for the Atlantis legend."

A 2002 nationally televised documentary by the BBC presented what was termed "fresh evidence": The Biblical plagues and the parting of the Red Sea were natural events rather than myths or miracles. The program suggested that the story of Exodus in the Bible can be explained as a result of the aftermath of the Thera eruption. Computer simulations of the effects of the Thera ash cloud and the generation of a tsunami are called on to explain several of the plagues and the parting of the shallow Sea of Reeds, the actual site of the exodus of the Jews out of Egypt, not the purported Red Sea.

The Thera eruption, regardless of whether it directly or indirectly caused the decline of the Minoan civilization, can be considered to be one of the most influential volcanic eruptions ever to have taken place in human history. While legends and myths are entertaining to the soul, scientific analysis of ash deposits, archeological digs, and the baseline provided by observations of presen-day volcanoes help reveal the real story about the great Thera eruption of the late Bronze Age. It was a catastrophic event that has and will continue to echo through time. Hopefully, with time, the echo will only get louder and clearer.

The baseline of scientific observations from historical volcanic eruptions reviewed in this chapter provides a working foundation for estimating the possible environmental and ecological effects of supervolcanoes to be discussed in subsequent chapters.

CHAPTER 4

Supervolcanoes

Double, double, toil and trouble; Fire burn and cauldron bubble.
—Macbeth, 4.1.10–12, William Shakespeare

The 2006 annual fall meeting of the American Geophysical Union (AGU), held in San Francisco from December 11 to 15, included a separate session on supereruptions. The session spanned two days and consisted of sixteen 15-minute oral presentations on Tuesday, December 12, and 26 poster presentations on Wednesday, December 13. The title of the session was "What Makes an Eruption 'Super'? New Methods Yield Insights About Very Large Calderas and Their Eruptive Products."

A brief session description preceding the compilation of program abstracts reads, in part, as follows:

The explosive eruption of large volumes of silicic magma can influence climate on a global scale, and has direct catastrophic effects, in some cases spanning continents. Recent television documentaries,

93

movies, and novels speculate on the initiation and aftermath of these rhyolitic super-eruptions, thereby focusing public attention on the current state of scientific knowledge. This session aims to summarize present understanding by inviting contributions on the histories of super-eruptions and the processes leading to large-volume ignimbrite sheets, and their atmospheric aftermaths.

The description then goes on to summarize detailed scientific lines of research that will be discussed during the two-day session, which will hopefully contribute to an understanding of the large-scale volcanic eruptions.

While many of the talks and posters presented in this two-day session are of special interest to the subject of this book, abstracts of the first two talks of the session, by leading researchers in this area of volcanism, are particularly noteworthy. The abstract for the first talk opens with the following sentences.

The evocative terms "supereruption" (and "supervolcano"), whilst eminently saleable to the media, conceal the fact that, apart from knowing that such large eruptions (>300 km^3, magma) actually have occurred, we understand very little about the dynamics of such events. Field studies of three supereruption deposits suggest that we are missing information on the timing and eruptive styles that is essential in assessing the dynamics and impacts of past and future large eruptions.

The abstract for the second talk opens with, "If the term super-eruption is to be adopted, then it needs to be defined. Recently it has been applied to large-scale, caldera and ignimbrite-forming explosive eruptions that have occurred in the geological past. The present suggestion is that it is reserved for eruptions that released >1 x 10^{15} kg of magma (roughly in excess of 500 cubic km of magma." The abstract concludes with the following sentences. "Volcanologists perhaps do not need the term super-eruption, but the term appears to catch the imagination of the public and media. The utility of the term supervolcano will also be discussed."

One of the notable aspects of the AGU session description and abstracts is that there is a remarkable variety in the spelling of supereruption or supervolcano. For instance, a tally of the spelling variations from the published abstracts is: *supervolcano* (five authors), *supereruption* (four authors), *super-volcano* (two authors), *super-eruption* (five authors; one of the five authors used *supervolcano*), and *super eruption* (six authors; one of the six used supervolcano). The spelling variations are, to some extent, a reflection of the relative youth of this aspect of the field of caldera eruptions. As the reader has already noticed, the spellings adopted in this book are supervolcano and supereruption.

The second point to be noted from the abstracts is that there is disagreement in the definition of a supervolcano in terms of volume of magma erupted. This disagreement is apparent when we consider that the USGS defines a supervolcano as an eruption of magnitude 8 on the VEI scale, implying, by their definition, that more than 240 cubic miles of magma are erupted. In this book, we adopt a magnitude scale proposed by three Earth scientists published in 2004 in the *Bulletin of Volcanology*. The new scale takes into account the density, or mass, of the erupted material rather than simply an estimate of the volume of ejected material or debris and, by so doing, mitigates inconsistencies that plague the VEI index scale. The inconsistencies arise when two volcanoes that eject similar volumes of debris are assigned similar VEIs, but one eruption consists largely of fluffy ash while the other eruption consists of dense volcanic rock. Obviously, this results in an inaccurate comparison of the lightweight eruption relative to the heavy-weight dense-rock eruption. We will return to the results of the study involving the new magnitude scale for a look at the temporal behavior of supereruptions following a discussion of their evolution.

The science of supervolcanoes is a relatively young science, and advances in knowledge are occurring on an almost monthly basis. Starting in the late 1940s when Lake Toba was identified as an extra large volcanic caldera, the potential sheer size and global effects of such events have commanded increasing attention from scientists of varying backgrounds. This is particularly evident given the range of scientific disciplines encompassed in the presentations at the fall 2006 AGU meeting.

ELEMENTS OF A SUPERVOLCANO

But what exactly is a supervolcano? A supervolcano is a particular type of explosive caldera eruption, namely a large-scale subsidence-resurgent caldera, where large-scale implies an exceedance of a threshold volume of about 120 cubic miles of material. The choice of 120 cubic miles will be explained later in this chapter. Before considering subsidence-resurgent calderas, we will discuss other fairly common types of caldera formations.

CALDERAS

In general, a caldera is a volcanic feature formed by the collapse of a surficial volcanic edifice into an underlying magma chamber that has been partially drained of magma, making it a large, special form of volcanic crater. The word *caldera* comes from a Spanish word meaning "cauldron."

Some volcanoes, such as Kilauea on the Big Island of Hawaii, form calderas in a different fashion. As we saw in Chapter 1, in the case of Kilauea, the magma feeding the volcano is relatively silica poor. As a result, the magma is much less viscous than the magma of a rhyolitic volcano, and the magma chamber is drained by large effusive lava flows rather than by explosive events. Such calderas are also known as subsidence calderas. The Kilauea Caldera has an inner crater known as Halema'uma'u, which has often been filled by a lava lake. The largest volcano on Earth, Mauna Loa, is also capped by a subsidence caldera called Mokuaweoweo Caldera.

Explosive craters or calderas are formed from the violent ejection of a significant portion of a stratovolcano. Such was the case for the Tambora volcanic eruption in 1815 and the Mazama eruption some 7,700 years ago (estimates vary from 7,000 to the age cited here) that resulted in Crater Lake in Oregon. While these events were considered devastating at the time of occurrence, they are still small compared to the supereruptions that are the result of the explosive formation of large-scale subsidence-resurgent calderas.

STAGES OF A SUPERVOLCANO

There are several stages that occur in the development and eruption of a supervolcano. The vast chambers of molten magma that give rise to supervolcanoes form either above hot spots (for example, Yellowstone) or above subduction zones (for example, Toba). In both cases, the supervolcanoes tend to follow a sequence of formative and eruptive stages, or an eruption cycle, that is better understood now than it was just a couple of decades ago.

While relatively recent advances have been made in the understanding of the eruption cycle of a supervolcano, it is abundantly clear from the content of the presentations at the December 2006 AGU meeting that considerable experimental (field and laboratory), theoretical, and modeling work remain. Two of the many factors that challenge the efforts of researchers in this area are the complexity of the Earth in terms of its material properties, and thermal regimes and the lack of direct observations or any kind of eyewitness accounts of a supereruption. The latter factor contrasts sharply with the situation for VEI 6 and smaller volcanic eruptions, wherein, ground- and space-based observations of actual eruptions are available for theoretical analysis and modeling. In the case of supervolcanoes, however, volcanologists often find themselves in a position similar to that of a cold-case homicide detective, who often has to start with very little information other than a suspicious death and has to uncover and piece together old and often partially preserved evidence to arrive at the facts of the case.

Our current understanding of a supereruption cycle is summarized in the following:

In the case of a subduction zone setting, the first stage of development involves partial melting of mantle rock above the sinking plate of oceanic crust. The resulting magma then works its way up toward the base of the continental crust and pools there. Eventually, the magma at the base of the crust begins to melt parts of the continental crust, which has a lower melting point than the rock below, forming the beginnings of a silica-rich magma pool.

The second stage involves the bulging or swelling, and the initiation of fracturing of the surface as a relatively shallow near-surface or upper magma chamber fills with magma from the lower chamber formed during the first stage at the base of crust. The upper magma chamber in the continental crust results in the accumulation of a silica-rich and relatively low temperature pool of molten magma formed from the mixing of the molten rock from below and the host rock. The higher silica content and lower temperature of the magma in the upper chamber is a more viscous, or flow-resistant, mix relative to the magma in the lower chamber. In the upper magma chamber, the molten material undergoes a chemical evolution. Less dense volatile materials are concentrated in the upper part of the chamber. These volatiles include the more silica-rich magma, various gases, and water. When ring fractures start to extend down from the surface to the upper chamber, the gun is loaded and cocked.

In the third stage as the magma chamber grows in volume, it stretches and bulges the crust above it. Because the upper crust is rigid and brittle, it fractures more easily than it bends. Thus, fractures, or faults, known as ring fractures,

Fig. 4-1. The four stages in the evolution of an explosive caldera supereruption:

1. *For more than several hundred thousand to millions of years, magma rises from deep in the Earth's crust; eventually the magma rises high enough and forces a dome to develop on the surface.*

2. *Magma pressure continues to increase, forcing ring fractures around the circumference of the domed area, from which vents begin to erupt, finally leading to an unzipping of the pending caldera circumference and a succession of major eruptions.*

3. *The eruptions exhaust themselves when pressure from below is relieved. That is when the caldera magma becomes relatively depleted. The roof caves in and a large-scale depression forms, which is gradually filled with rainwater.*

4. *Magma pressure begins to rise again from below, and an island is pushed up in the lake. Rich folks buy property on the secluded island and build their palatial mansions, never realizing that their property could end up in the stratosphere where real estate values are nonexistent.*

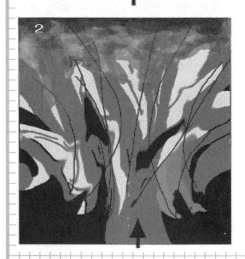

Figure and descriptive text is courtesy of George Weber from his online book Toba Volcano available at www.andaman.org/BOOK/originals/ Weber-Toba.

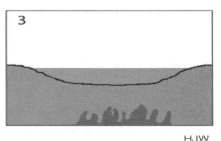

HJW

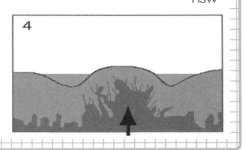

develop along the circumference of the bulge. As the bulge grows, the ring fractures propagate downward, eventually intersecting the primed upper-magma chamber. Gases expand in the magma, making a much larger volume of frothy fluid. The expanding, low-density, hot gases and magma mixture rise rapidly along the ring fractures and vent at the surface in the form of sustained explosions of white-hot froth. Driven by the hot gases, giant fountains of in-candescent ash burst from the ring fractures at temperatures near 1,800°F. Plumes of ash are jettisoned into the stratosphere where planetary winds carry them around the world, eventually blanketing tens of thousands of square miles with volcanic ash. Nearer to the vents, fiery clouds of dense ash, fluid-ized by the expanding gas, boil over caldera rims and rush across the country-side at speeds of more than 100 miles per hour, vaporizing all plant and animal life in their path. Gaping ring fractures that have extended downward into the upper magma chamber provide conduits for continuing foaming ash flows. Fractured pieces of rock comprising the volcanic edifice plunge down into the chamber, forcing additional magma up the outside edges of the ring. The col-lapse may occur as the result of a single massive eruption, or it may occur in stages as the result of a series of eruptions. The upper chamber generally collapses before more than one tenth of its magma is drained.

During the fourth and final stage after the eruption has ceased, a craterlike depression, or caldera, initially sits above the partially drained magma cham-ber. Gradually, the land within the caldera begins to dome up as magma moves back into the upper chamber, forming a resurgent caldera and potentially ini-tiating another supereruption cycle. In many locations around the world, the caldera is filled with rainwater, forming a deceptively beautiful, serene lake such as Lake Toba.

The following picture is an aerial view of Crater Lake looking east. This beautiful, deep blue lake, located in south central Oregon, was the site of a massive eruption and collapse of a mountain known as Mount Mazama about 7,700 years ago (as noted previously, some sources quote 7,000 years ago) that resulted in a classic caldera formation about 4,000 feet deep and 5 by 6 miles in aerial dimensions. While not a supervolcano per se, with "only" roughly 12 cubic miles of ejected material, it is included here because it's an excellent example of a caldera with the formation of a central platform, volcanic cone

Fig. 4-2. Aerial view from the west of Crater Lake with Wizard Island in the foreground on the western edge of the caldera floor just sticking its peak above the crater rim. USGS Photograph taken by Mike Doukas on December 10, 2005.

(Wizard Island, Merriam Cone), and other smaller volcanic features, including a dome that was eventually created atop the central platform. Unlike many supervolcanoes, Crater Lake is small enough to be captured in one aerial photograph.

Crater Lake, the result of rainfall and snowmelt, is the deepest lake in the United States, second deepest in North America (Great Slave Lake in Canada being the deepest), and the seventh deepest in the world. It also holds the honor of being the deepest lake in the world that is completely above sea level. A maximum lake depth of 1,996 feet was recorded by the USGS in 1886 using piano wire and a lead weight. A more recent measurement in 1959 by the USGS using sonar techniques yielded a maximum depth of 1,932 feet.

Resurgent calderas, among the largest volcanic structures on Earth, are associated with massive eruptions of voluminous pyroclastic sheet flows, on a scale not yet observed in historic times. These large-scale eruptions can produce hundreds or even thousands of cubic miles of material. A period of explosive volcanism rocked the western United States approximately 25 to 40 million years ago, producing hot ash flows and volcanic deposits that covered extensive parts of the southwest region. This event, termed the Mid-Tertiary Ignimbrite Flare-up (MTIF), erupted an awesome amount of igneous material, estimated at more than 120,000 cubic miles of rock! Such an astounding magnitude of material is hard to imagine. One way to better comprehend what this meant is to consider that the state of Colorado has a land area of about 104,000 square miles. If all the material, rock, and ash, ejected during the flare-up was gathered up and dumped on Colorado, a blanket of rock more than a mile thick would cover the state! Another way to look at it is that the MTIF was almost 200,000 times as voluminous as the 1980 eruption of the familiar Mt. St. Helens volcano.

The MTIF is famous, or infamous, for another reason. One of the largest known supervolcanoes that ever occurred on the Earth erupted during the flare-up. This supervolcano is in a class of its own.

LA GARITA

Located in the San Juan Mountains in southwestern Colorado is the La Garita volcanic caldera, one of a number of calderas that formed in the states of Colorado, Utah, and Nevada during the MTIF 25 to 40 million years ago. La Garita was the site of a truly enormous eruption about 26 to 28 million years ago. This was the Fish Canyon Tuff, with an estimated volume of approximately 1,200 cubic miles. The area devastated by the eruption covered a significant portion of what is now Colorado, and ash is thought to have fallen as far as the east coast of North America and the Caribbean.

Because of the large size of the La Garita eruption and the erosion of the remnants of the eruption over the past approximately 28 million years, it took geologists nearly 30 years to first appreciate and then determine the enormity of the resulting caldera. Modern-day estimates of the size of the caldera make it approximately 22 x 47 miles, giving it an unusually oblong shape. Truly, La Garita belongs in the supervolcano family at the head of the class. Fortunately, for a good portion of the population of the United States, it is recognized as an extinct supervolcano, rather than a dormant one.

There are three resurgent calderas in the United States less than 1.5 million years old: the Valles Caldera in New Mexico, the Long Valley Caldera in California, and the Yellowstone Caldera in Wyoming. Unlike La Garita, the Long Valley and Yellowstone calderas continue to show signs of present-day unrest.

One of the astonishing recent findings about a supervolcano has to do with the estimated duration of the explosive phase that results in the formation of a caldera. As an example, the Long Valley supervolcano ejected between 120 and 160 cubic miles of magma some 760,000 years ago. Remnants of the ejected material are seen today in a volcanic layer tens to hundreds of feet thick in eastern California known as the Bishop Tuff. For many years, geologists thought that the Bishop Tuff, given its extent and thickness, must have been formed by a series of eruptions over millions of years. However, as a result of laboratory analyses and field studies carried out in the 1990s, geologists now believe that the Bishop Tuff was ejected in the Long Valley supereruption lasting, not millions of years or even years, but a mere 6 days. Amazingly, the Toba supereruption approximately 74,000 years ago, which covered nearly 1 percent of the Earth's surface with some measure of ash, is estimated to have had a duration of only 9 to 14 days.

As is the case for many aspects of the Earth's behavior, however, complexity reigns. It was recently reported that the durations of the supereruptions at Taupo, New Zealand, approximately 26,500 years ago, and at Yellowstone about 2,100,000 years ago were on the order of weeks in the case of Taupo to as much as a year or more for Yellowstone. When compared to Long Valley and Toba, it is obvious that there is no simple, or cookie-cutter, model that can be exercised to predict the eruptive history of all supereruptions. This caution applies to other aspects of eruption dynamics as well, such as development of ash falls and flows, and implies that in the event of a future supereruption, assessing the end of the eruption, or foreseeing climatic stages that might have the greatest impact, could be problematic at best. This complicates hazard and risk analyses and could play havoc with any attempts at immediate emergency efforts in the vicinity of a supereruption.

A scenario that has recently been proposed for a relatively short-duration supereruption on the scale of Long Valley or Toba would feature blasts of superheated, foam-like gas and ash that rise buoyantly into the Earth's stratosphere, some 20 to 30 miles high. As the land above the magma chamber collapses into the partially depleted chamber, immense pyroclastic flows burst out along the ring fractures around the caldera at speeds in excess of several hundred miles per hour. The flows burst out at temperatures of 1,100 to 1,300°F, and burn and bury everything for tens of miles in every direction.

It's in the atmosphere, however, that the supereruption has even more spatially and temporally far-reaching consequences. When supervolcanoes blow, they can cover entire continents with ash. How this happens has been a bit of a mystery, however, given that wind and the initial estimated force of the eruption are not enough to carry the ash over such long distances. Pale-gray ash injected into the atmosphere would fall like clumps of snow for hundreds of miles for days or weeks. Examination of prehistoric eruptions by scientists from Australia and the United Kingdom has supplied a possible solution. Using geological records of ash volume and magma chamber size, the scientists estimated the energy of the prehistoric blasts and the plumes they would have generated. What they found was that the Earth's rotation literally fans the ash in the plume into a giant spinning cloud up to 3,700 miles wide within one day. To quote one of the scientists, "It's a bit like a hurricane, but on a much larger scale."

In Chapter 9, the devastating effects of ashfall on ancient mammals in Nebraska from a volcano approximately 1,000 miles away in Idaho are described. The estimated volume of magma ejected from this event was only about 60 cubic miles, making it less than twice the size of Tambora and considerably smaller than what we think of a supervolcano, and yet it wreaked havoc on a variety of mammals over a period of a few days to weeks. One can only imagine what the "downwind" impact of an actual supervolcano would be!

As mentioned earlier, a new magnitude scaling system that is based on estimates of erupted mass was published in 2004, and applied to 47 of the largest explosive eruptions that occurred in the past 500 million years, with 42 of the 47 events occurring in the past 36 million years. In the 2004 study, the largest events are defined to be those eruptions that produce more than 10^{15}

kilograms of material, which corresponds to a bulk volume of 120 cubic miles of material. This definition is adopted in this book as the definition of a supervolcano.

Most of the 2004 analysis focuses on the 42 events in the last 36 million years.

The following table is a partial listing of the revised magnitudes for several of the supereruptions that have been discussed. The magnitude of an event is based on a simple logarithmic relation involving the mass of erupted material. A minimum and maximum magnitude is computed for each event where the maximum magnitude is calculated based on an adjusted eruption mass determined in the 2004 study. The Toba and Taupo supereruptions are two of only three events that had sufficient field data available, so as not to require a calculation for an adjusted mass, resulting in equal minimum and maximum magnitude estimates. Notice that the La Garita caldera is the only event with a minimum magnitude in the M9, range and that events with bulk volumes as small as 120 cubic miles (for example, Long Valley) are considered supereruptions.

Caldera Name	Bulk Volume (cubic miles)	Age (million years)	Magnitude	
			min	max
La Garita	1,200	27.8	9.1	9.2
Toba	670	0.074	8.8	8.8
Yellowstone	590	2.0	8.7	8.9
Paintbrush Tuff Topopah Spring Tuff	290	12.8	8.5	8.9
Taupo	280	0.0265	8.1	8.1
Long Valley	120	0.7	8.1	8.5

The bulk volume for Taupo, 280 cubic miles, is based on material with a considerable amount of void spaces. The dense rock equivalent volume for Taupo is substantially reduced to a value of approximately 125 cubic miles, more in line with the dense rock equivalent volume of 110 cubic miles for the Long Valley caldera eruption.

An interesting event listed in the table is the Paintbrush Tuff. The Paintbrush Tuff refers to the group of tuffs, which includes the Topopah Spring Tuff, comprising Yucca Mountain, located in Nye County, Nevada, about 90 miles northwest of Las Vegas. Yucca Mountain derives its fame from the fact that it is the proposed site of the nation's repository for spent nuclear fuel and high-level radioactive waste. The repository host rock is the Topopah Spring Tuff that was formed 12.8 million years ago (Yucca Mountain Site Description, 2004) as the result of a large explosive caldera eruption, most likely a supereruption, to the north of Yucca Mountain. The large-scale volcanic eruptions that produced Yucca Mountain ended about 11 million years ago.

Then several million years ago, a different type of volcanic activity began in the area. Now instead of massive explosive eruptions like those that formed Yucca Mountain, the new activity has been characterized by relatively small and much less explosive eruptions that resulted in the formation of basaltic scoria cones. The last small eruption occurred between 75 and 80,000 years ago near Lathrop Wells, Nevada, about 12 miles south of Yucca Mountain.

A group of 10 experts concluded in a report published in 1996 that the chance of a volcanic event disrupting the repository is about one in 63 million per year. A new expert elicitation is underway and will include data collected since 1996, especially age dates for samples obtained from boreholes recently drilled in several key locations. Final results are expected in 2008.

Analyses of the temporal behavior of the 42 largest eruptions in the past 36 million years reveals some interesting patterns. First, the supereruptions are not spread out uniformly throughout the 36 million years. Instead, the events are clustered in two pulses or time periods of greatly increased globally distributed activity. While the available record of supereruptions is considered incomplete, the pulses are fairly robust and considered to be real. The pulses, lasting from about 36 to 25 Ma, and 13.5 Ma to present, approximately coincide with increased rates of deep-sea volcanic ash sedimentation and principal global climate cooling steps in the Cenozoic era spanning the past 65.5 million years. The pulse of activity covering the period from 36 to 25 Ma encompasses the most active phase of the San Juan volcanic field in Colorado and includes a total of 22 M8 eruptions and one M9 eruption, yielding a minimum supereruption frequency of 2 events/Ma. The present pulse, or flare-up, includes 19 M8 eruptions in the last 13.5 Ma for a minimum supereruption frequency of ~1.4 events/Ma (or 2 events/Ma for the past 6 Ma). The clustered nature of these supereruptions adds weight to suggestions put forward by several earth scientists that the rate of occurrence of these events is non-uniform, consisting of discrete packets of increased activity that are presumably controlled by regional or global-scale tectonic changes.

Application of a statistical approach involving Extreme Value Theory to the limited data set of supereruptions yields lower and upper bound estimates for the size-frequency relationships of eruptions of M8 and larger. The lower bound frequencies for M8 and M9 eruptions are 1.4 and $<10^{-4}$ events/Ma, respectively. Corresponding upper bound frequencies are 22 events/Ma for M8 eruptions and 0.15 events/Ma for M9 eruptions. The 22 events/Ma, or 1 event per 45,000 years is close to previously published estimates of supereruption frequencies. Based on the pulse of activity during the past 13.5 Ma, statistical analysis yields a 75 percent probability of a M8 supereruption occurring within the next million years and a 1 percent chance of such an event in the next 460 to 7,200 years. However, given the suspected incompleteness of the record, these estimates should only be considered as minimum values. With new field and laboratory studies of known, or suspected, large silicic volcanic provinces that, to date, have only been poorly studied, it is highly likely that the number of M8, and possibly M9, supereruptions will increase, and perhaps significantly. One indication of this is a 2006 report by Argentinean geologists wherein the scientists described the uncovering of what is variously termed a "herd" or "gaggle" of supervolcanoes hidden in the remote Argentine-Bolivia-Chile highlands. One of the supervolcanoes in particular, the Vilama Caldera, may have matched or exceeded the explosive fury and pyroclastic volume of the Yellowstone eruptions. It's estimated that the Vilama Caldera ejected some 500 cubic miles of magma in a single gigantic eruption approximately 8.4 million years ago.

The availibility of rates of occurrance of large explosive eruptions discussed above allows a direct comparison of the hazard of these events with other large magnitude natural phenomena. The other natural phenomena include asteroids impacts, earthquakes, and hurricanes. Of the phenomena mentioned, only asteroid impacts come close to posing a hazard comparable to supereruptions. When the energy release and frequency of occurrence of impacts and large eruptions are compared, the available data suggest that on timescales of about 100,000 years or less volcanic eruptions in the magnitude range M8 to M9 result in a greater hazard to mankind and civilization. If future studies results in an increase in the number of M8 and larger eruptions, especially during the last pulse of activity between 13.5 Ma to the present, then the hazard associated with these events would increase accordingly.

Simple linear extrapolation of the regional and global environmental effects of the historic and prehistoric volcanoes to the scale of an M8 to M9 supereruption predicts dire consequences for civilization here on Earth. While linear extrapolation is questionable, it is clear that the global effects of the ash fallout and aerosol clouds on climate, agriculture, health, and world economies and governments would pose a severe challenge to modern civilization. In the following chapters, we'll address the threat to civilization that occurred about 74,000 years ago when the Toba caldera exploded and, in a later chapter, consider the modern-day threat from a hypothetical eruption of the Long Valley, California, supervolcano.

PART II

Toba

CHAPTER 5

Civilization exists by geological consent, subject to change without notice.

—Will Durant, American Writer and Historian

Beneath the Indonesian Archipelago, the Earth never sleeps. As part of the Ring of Fire, this volatile region, made up of more than 18,000 islands extending almost 3,200 miles east to west between the Indian and Pacific Oceans, experiences about 75 percent of all worldwide volcanic activity occurring on the earth's surface, with more than 130 active volcanoes. And when the sky isn't being filled with the eruptive gasses and ash clouds, and the hills and calderas are not flowing with lava, earthquakes and tsunamis threaten the existence of the area's 221 million inhabitants.

The largest islands in the Archipelago, or chain of islands, often volcanic in nature, are Sumatra, Java, Sulawesi, Kalimantan, and part of New Guinea. The Indonesian Archipelago is the home of the Spice Islands, a critical trade destination since ancient times, and is the world's largest archipelagic state, the capital of which is Jakarta. It is a region of both geological and political turmoil, as well as a favorite tourist destination for those who want a taste of the exotic.

Most people know the region, however, for its catastrophic natural disasters, from the eruptions of Tambora in 1815 and Krakatoa in 1883, to the devastating earthquake and tsunami in December 2004 that took hundreds of thousands of lives. As pointed out in Chapter 3, the 1883 Krakatoa eruption is the subject of a 1969 Hollywood film starring Maximilian Schell, which was titled *Krakatoa, East of Java*—even though Krakatoa is in fact west of Java. This blatant error is perhaps the most remembered thing about the film. Tambora, on the Island of Sumbawa, is the violent volcano east of Java.

Few people realize that this land of major tectonic activity is also rich in history. It is home to half of the world's hominid fossils, found at Sangiran in Java. This was also the home of "Java Man," existing approximately 500,000 to 2 million years ago, and is the home of the 3-foot-tall miniature hominid "Flores Man," discovered recently on the island of Flores.

Sometimes, history and catastrophe become one and the same, as in the case of a serene and lush lake on the island of Sumatra named Toba.

SUMATRA

Sumatra is the largest island located entirely in Indonesia with a total population of more than 40 million. Its name in Sanskrit means "Isle of Gold," referring to mines in the highlands that once exported gold. Sumatra suffered massive devastation from the 2004 magnitude M 9.3 earthquake and tsunami. In addition to the subduction zone and the associated Sunda Arc off the west coast of the island, Sumatra is also the site of a large transform fault, the so-called Great Sumatran Fault, also called the Sumatra Fracture Zone, running the entire length of the island. Although an earthquake on this fault will not create a tsunami, it will still probably have disastrous consequences, due to its proximity to major population centers. For instance, the northwestern end of the fault lies directly under the famous city of Banda Aceh, which was devastated, with more than 30,000 people killed, by the tsunami generated by the December 2004 Indian Ocean earthquake. The city's fame comes from the fact that Islam first arrived there (Bandar is Persian for *port* or *haven*) and, from there, spread throughout Southeast Asia.

The Aceh area is considered to be a religiously conservative Islamic society and has had neither tourism nor any Western presence in recent years due to armed conflict between the Indonesian military and Acehnese separatists. Some believe that the tsunami was punishment for lay Muslims shirking their daily prayers and/or following a materialistic lifestyle. Others have said that Allah was angry that there were Muslims killing other Muslims in an ongoing conflict. As said by Ebiet G. Ade, an Indonesian ballad singer, in his song, *Berita Kepada Kawan* (News for Friend):

Mungkin Tuhan mulai bosan melihat tingkah kita/yang selalu salah dan bangga dengan dosa-dosa/atau alam mulai enggan bersahabat dengan kita.

(Maybe God begins to be bored watching our behaviors/that are always wrong and proud with the sins/or nature begins to be reluctant to be friendly with us.)

In what may be the most significant positive result of the tsunami, the widespread devastation led the main rebel group Geraken Aceh Merdeka (GAM) to declare a cease-fire on December 28, 2004, shortly followed by the Indonesian government. The two groups resumed long-stalled peace talks, which resulted in a peace agreement signed August 15, 2005. This agreement explicitly cites the tsunami as a justification.

Sumatra boasts a host of flora and fauna that take advantage of the rich volcanic soil, especially along the shores of Lake Toba, or Danau Toba. About 62 miles long and 18 miles wide, the lake sits in the middle of the northernmost part of Sumatra. Along its fertile banks, the people, most of them of the ethnic group Batak, build their traditional Batak houses with their curved, bow-like roofs. The Bataks today spend their time weaving, wood carving, and creating ornate stone tombs. They have a rich history of religious tradition, mostly Christian but with a Muslim minority and speak a variety of Austronesian languages. Their lives are simple, living beside the serene lake.

One has to wonder if they are collectively aware of the catastrophic event that created the very lake they live beside each day, or if they know that the ground beneath their feet is most likely one of the largest resurgent calderas on Earth. Do they know that their lovely lake exists near a fault line known as the Great Sumatran Fault, and that their volcanoes are part of the Sunda Arc, a subduction zone described as one of the most tectonically active zones in the world?

TOBA

Do the people boating and fishing along the shores of Lake Toba today realize what happened there approximately 74,000 years ago, when a single event changed the entire course of human evolution, and reshaped the very land they have come to call home?

Long ago, Lake Toba was far from serene. It was, instead, a monster.

The story leading up to the recognition of Lake Toba as the site of one of the largest volcanic eruptions, a supervolcano, in perhaps the last 27 million years begins in the mid- to late-1940s. Following World War II, the Dutch government assigned the Dutch geologist Reinout Willem van Bemmelen the job of recollecting all information on the geology of the Indonesian Archipelago lost during the war. In 1949, he published a book *The Geology of Indonesia*, which included results of his field work in the Lake Toba region. Van Bemmelen reported that the lake was surrounded by a vast layer of ignimbrite rocks, which, as we recall from earlier discussions, is welded tuff. Lake Toba was apparently the scar or caldera of a huge volcano! Just how huge in terms of the amount of material ejected and the short- and long-term aftereffects on the global environment and humanity would be pieced together over the next several decades by scientists of widely differing scientific disciplines.

Van Bemmelen in 1939, and other scientists as recently as 1984, suggested that Lake Toba was formed by the filling of a depression that was left behind by a single, large eruption. Subsequent studies showed, however, that the depression filled by Lake Toba was the result of four episodes of explosive eruptions and caldera collapses. These studies included age dating and stratigraphic analyses of samples of ash-flow tuffs from each of the four eruptive units, and a tomographic analysis of seismic travel-time data for determination of the three-dimensional P-wave velocity structure beneath the caldera complex.

A comprehensive description of the evolution of the Toba caldera complex is laid out in a paper published in the *Journal of Petrology* in 1998 by Craig Chesner. The paper is the result of investigations of the petrology (chemical and mineralogical composition) and stratigraphy (sequence of layering) of the ash flow tuffs that comprise the caldera complex. The eruptive history of the

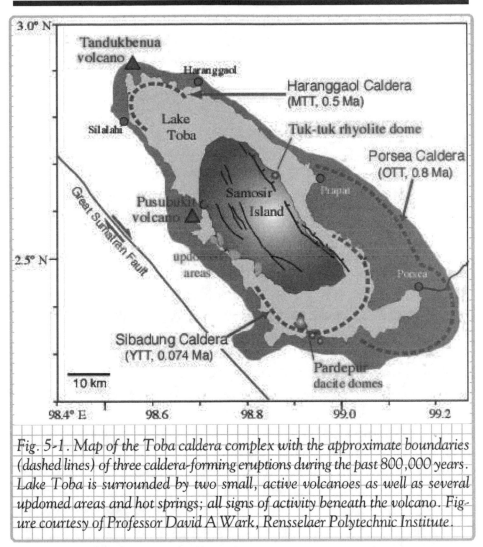

Fig. 5-1. Map of the Toba caldera complex with the approximate boundaries (dashed lines) of three caldera-forming eruptions during the past 800,000 years. Lake Toba is surrounded by two small, active volcanoes as well as several updomed areas and hot springs; all signs of activity beneath the volcano. Figure courtesy of Professor David A Wark, Rensselaer Polytechnic Institute.

Toba complex began more than a million years ago with the formation of a large stratovolcano within the northern boundaries of the present caldera (see Figure 5-1). Then, about 1.2 million years ago, the stratovolcano erupted an estimated 8 cubic miles of material. This eruptive unit is referred to as the Haranggaol Dacite Tuff (HDT)—named Haranggaol for its proximity to the lakeside village of the same name, and dacite for the dominant type of volcanic rock particular to this eruptive unit. Dacite is intermediate in composition between andesite and rhyolite. The size of this eruption is comparable to revised estimates of the volume of the 1815 Tambora eruption, 7 to 8 cubic miles of magma, published in 2004 in Geophysical Research Letters. At 8 cubic miles, the HDT eruption is a VEI 6 event, considerably smaller than two of the supereruptions to follow.

The next event in the Toba sequence occurred in the southern half of the caldera complex about 800,000 years ago. In Figure 5-1, the caldera formed by this event is referred to as the Porsea caldera. The estimated volume was 120 cubic miles, putting it in the supervolcano category according to the criterion (bulk volume ≥ 120 cubic miles) described in Chapter 4. This supereruption was the first known quartz-bearing Toba tuff, and is named the Old Toba Tuff (OTT) in the literature.

About 500,000 years ago, activity switched back to the northern portion of Toba with the eruption of the quartz-bearing unit referred to as the Middle Toba Tuff (MTT). The estimated volume of this event was 14 cubic miles, making it a VEI 6 event, again considerably smaller than the supervolcano that would follow. The location of this eruption is represented by the dashed contour forming a half circle, and is designated the Haranggaol Caldera in Figure 5-1. The MTT eruption possibly originated from the same caldera that resulted in the HDT event 1.2 million years ago.

Finally, 74,000 years ago, the unit referred to as the Youngest Toba Tuff (YTT) erupted in what was possibly the largest explosive event since the La Garita Caldera eruption of the Fish Canyon Tuff in Colorado nearly 28 million years ago. The estimated eruption volume of the YTT event is 670 cubic miles. The southern half of the caldera resulting from the YTT supereruption is noted in Figure 5-1 as the Sibadung Caldera. Scientists disagree on the northern extent of the YTT caldera. One opinion is that the caldera extends all the way north to include the HDT location, and the other view is that the YTT caldera follows the boundary of Samosir Island, but does not extend any further north.

Several researchers have commented that the repose time or period between the four eruptions within the Toba caldera complex is between 0.34 and 0.43 million years. While this is the case for the past four eruptions, there are two caveats that we add to this statement. First, the stated interval of repose periods is based on a statistically small sample, and may not be meaningful in terms of the timing of the next eruption within the caldera complex. Second, the repose period between supereruptions is on the order of 750,000 to 800,000 years, but only based on two events. The main point here is that the two relatively small events at 1.2 and 0.5 million years ago should not be included in the sample for determining repose periods of the supereruptions. As we will see in Chapter 9, similar comments apply to the sequence of eruptions at the Yellowstone caldera, and could imply that the so-called "overdue" supereruption at Yellowstone is in fact not overdue at all.

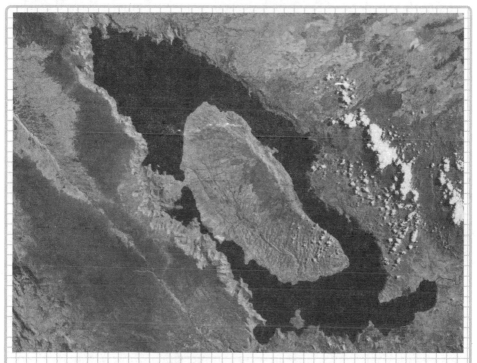

Fig. 5-2. Landsat satellite photo of Lake Toba in northern Sumatra, Indonesia. Resurgent doming of the Toba caldera formed the prominent island of Samosir. This island is connected to the mainland by a narrow isthmus on its west side. The Great Sumatran Fault can be seen cutting across the lower left-hand corner of the photo in a northwest-southeast direction. Image courtesy of NASA.

Evidence that the caldera is still an active magmatic system comes from the existence of hot springs along the western caldera rim, by the post-74,000 year resurgence of the caldera floor as seen by Samosir Island, and by the existence of the young (post-YTT) volcanoes Tandukbenua and Pusubukit. From the regional tectonics, we note that the position of Toba, along the Sunda arc and above a subducting slab, is consistent with the assertion that magma still occupies the crust below the caldera complex.

The 1.2-million-year-old Toba complex is one of the largest resurgent calderas on Earth, approximately 62 miles long by 19 miles wide. Following the YTT eruption 74,000 years ago, magma started flowing back into the partially depleted chamber, and gradually pushed up the land above the surface of the lake to form what is now the island of Samosir and the Uluan Peninsula to the east of Samosir. Lake sediments on Samosir show that it has been uplifted by at least 1,476 feet. The island was originally connected to the caldera wall on its west side by a small isthmus, which was cut through to aid navigation. In keeping with the record size of the Toba caldera, Samosir Island also sports some hefty numbers. At 243 square miles, Samosir is notable for being the largest island within an island (Sumatra), and the fourth largest lake island in the world.

The false image Landsat photo of Lake Toba in Figure 5-2 shows all the features discussed here in dramatic detail; in particular, the resurgent island of Samosir, the isthmus connecting Samosir to the mainland, the Uluan Peninsula, and a very obvious fault trace. The Great Sumatran Fault, a right-lateral strike-slip transform fault zone, cuts across the lower left-hand corner of the photo in a northwest-southeast direction. Examination of the photo reveals some features that appear to be offset in a right-lateral direction.

A study that provides an estimate of the three-dimensional velocity structure of the crust and upper mantle beneath the Toba caldera complex was published in 2001 by scientists from the United States and Indonesia. The study consisted of tomographic analysis of travel-time data obtained from the arrival times of P-waves (first arriving compressional waves) from locally occurring earthquakes recorded by a network of 40 seismic recording stations. The network operated in the region for four months and recorded approximately 1,500 local earthquakes originating in the subducting slab beneath Toba, from the Sumatra Fault Zone, and from faults within the Toba depression.

Seismic tomography is analogous to X ray tomography in medicine, more commonly known as a CAT scan. Instead of X rays passing through a body from different angles, seismologists use P-waves from earthquakes at different distances and azimuths from the volume of material to construct a three-dimensional image. A basic assumption invoked in the interpretation of results from seismic tomographic analyses, especially in volcanic settings, is that areas through which waves move quickly tend to be cooler or consist of denser rock. However, areas where P-waves move slowly indicate warmer or less dense rock, such as magma.

One of the major results of the study is that there is a dominant low-velocity region in the upper 6 miles of the crust, possibly encompassing an area as large as the southern two thirds of the present-day 62×19-mile caldera complex. In the middle crust, from 6 to 15 miles depth, this main low-velocity volume narrows in size and appears to merge with a more focused, low-velocity region below the Pusubukit volcano (Figure 5-1). This region of low velocities is even more focused in the lower crust 12 to 24 miles, and can be traced into the uppermost mantle. Thus, what is seen at depth is a complex structure that formed by coalescence of smaller calderas. The large region of low velocity is separated by a high-velocity zone from a second low-velocity region that underlies the northern end of Toba. Smaller in size than that which underlies the southern 2/3 of the Toba caldera complex, the northern low velocity region coincides with the location inferred for the MTT caldera-forming eruption 500,000 years ago. The main, southern low-velocity region, which coincides with the site of resurgent doming, outlines the reservoir that fed the large ash flow eruption 800,000 years ago, and again 74,000 years ago during the eruption of the YTT supervolcano.

The finding of two separate chambers disagrees with the result of an earlier study, which asserted that the 62-mile-long caldera complex is underlain by a single, shallow magma body. However, the authors of the tomographic study added the caveat, that they were not able to rule out the possibility that a single magma system existed in the past, or that the present system is connected at depths where the spatial resolution of the seismic data is low.

TOBA'S EFFECTS

We now turn to the cataclysmic effects that the YTT Toba supereruption, 74,000 years ago, had on the global environment and the human species. The Toba event left behind many clues in the geological record that scientists have pieced together to arrive at an appreciation and amazement of the enormity of this volcanic eruption.

On September 26, 2006, PBS aired a program titled *Mystery of the Megavolcano*. During the course of the program, seven Earth scientists, with different scientific backgrounds and specialties, told how they each uncovered a clue that pointed to something huge that had happened 74,000 years ago. And when they put the different clues in a single pot and stirred, out came the "eureka moment" scientists revel in, when their persistent efforts finally paid off.

One of the clues that contributed to the solution of the mystery was a very large electrical conductivity spike in a Greenland ice core dated at approximately 75,000 years. As we saw in Chapter 3, a spike in electrical conductivity is caused by the presence of sulfuric acid in an ice core, and is generally interpreted as coming from a volcanic eruption. However, the large amplitude of the signal in this case meant that the exceptionally high concentrations of sulfuric acid were caused by something that Greg Zielinski, the scientist conducting the ice core study, could only describe as "cataclysmic." The PBS narrator commented that the quantity of sulfuric acid released by the event was "up to 25 times more than all industrial sulfuric acid pollution over an entire year on Earth today." The presence of the acid in such high concentrations implied that the causative mechanism was most likely not an impact by an extraterrestrial object, because there is no evidence of impacts producing sulfuric acid. While volcanic eruptions are known to produce sulfuric acid (see Chapter 3), the amount of acid contained in the ice core meant that if indeed the source was a volcano, it would have to have been thousands of times more powerful than any events in historic times. But while the Greenland ice core pointed to something cataclysmic happening to the Earth about 75,000 years ago, possibly a volcano, it did not indicate where.

The next clue came from a core drilled into the ocean floor thousands of miles from Greenland. Geologist Michael Rampino, an expert in volcanology and deciphering ocean cores for the Earth's climatic history, came upon a sudden drop in ocean temperatures that occurred about 75,000 years ago. His analysis was based on the relative abundances of two oxygen isotopes contained in the shells of tiny ocean creatures called foraminifera. He estimated the magnitude and rate of the temperature drop at about 10°F or more in as little as a few thousand years—a geological blink! In his words: "It was like flipping the switch on the global climate system from hot to cold." While the answer to where the event had occurred would wait on the results from another study, the scientists were closing in on what it was. The evidence was favoring a volcanic eruption, but one of unheralded proportions.

The stage was set for the entrance of John Westgate, a quaternary tephrochronologist from the University of Toronto. His specialty is identifying volcanoes based on the chemical composition of the ash that they eject and scatter about. His encounter with Toba began in 1990 when people started to send him ash samples from widely separated parts of the Earth. The surprising and, at the same time, puzzling thing was that all the samples were very similar in chemical composition. It was surprising because these samples were collected at locations differing by 3,700 to 4,300 miles, and yet the chemistry was pointing to a single source. His next step was to date some of the samples. A process known as fission track dating yielded the same age for all the samples tested: 75,000 years old. Now all he had to do was find the volcano.

In Chapter 3, the environmental impacts of the Icelandic eruption, Laki, were described. Given its proximity to Greenland and the possibility that the relatively close distance might implicate an ancient Icelandic volcano in the ice core that Greg Zielinski had analyzed, the chemical compositions of ash samples from the Laki volcano, playing the role of proxy samples for a possible prehistoric eruption, were determined. Westgate's analyses showed, however, that none of the Laki ash samples matched the chemical composition of his mystery samples. Finally, in the spring of 1994, he received an ash sample sent by Craig Chesner from a new location in Southeast Asia. The sample came from a tropical jungle on the Indonesian island of Sumatra on the shores of a beautiful, serene lake: Lake Toba. It didn't take long before the answer was in. The sample from Lake Toba matched the chemistry of the mystery samples.

The next and final clue came from a survey of the geology of the Lake Toba region and a bathymetric survey of the lake. Chesner found an incredibly steep profile for the lake and the cliffs rising up out of the lake for hundreds of feet (refer to the Lake Toba photo on page 14 and note how steep the caldera walls are). While most lake bottoms drop off gradually from the shore, that's not the case for Lake Toba. Just a short distance from the shore, he was recording depths as large as 575 feet. Then, when he climbed the steep cliffs above the shoreline, Craig found rocks characteristic of a volcanic eruption extending all the way to the top. This represents a tremendous amount of magma, thousands of times greater than what is found for the scale of volcanoes considered in Chapter 3.

Thus, the pot was stirred and the answer came out. Evidence from ice cores, ancient sea shells, ash samples, and geologic and bathymetric surveys all pointed to an explosive volcanic eruption of enormous proportions.

The volume of material erupted from the Toba YTT was noted as 670 cubic miles earlier in this chapter. That estimate consists of 480 cubic miles of pyroclastic tuff flow deposits, and a widespread ash layer of approximately 190 cubic miles. The eruption cloud produced by Toba has been estimated to have reached heights of more than approximately 20 miles, well into the strato-sphere. Somewhat surprisingly, the duration of the eruption based on ash falls in the Indian Ocean has been estimated at two weeks or less. Using a two-week eruption duration for YTT and an eruption volume of 670 cubic miles results in an absolutely amazing average eruption rate of about 8 million metric tons of material per second.

As discussed in Chapter 3, the amount of sulfur volatiles released during a volcanic eruption is extremely important in terms of the magnitude of the cli-matic impact of the event. In the case of Toba, estimates of the amount of sulfuric acid aerosols, based on studies of the sulfur content of the supereruption deposits and on data from smaller historical explosive eruptions, range from 1,000 Mt to 5,000 Mt.

The enormity of Toba in all aspects prompts two questions: What impact did this eruption have on the global climate, and how did humans alive at that time cope with such a cataclysmic event? Michael Rampino and Stanley Ambrose made a very important observation in a Geological Society of America report published in 2000 (Special Paper 345) that must be kept in mind as we try to answer the questions posed in the preceding sentence. To paraphrase, the climatic and environmental impacts of the Toba supereruption are so much greater than those following historical eruptions that anecdotal information, instrumental records, and climate-model studies of the effects of historical eruptions may not be relevant in scaling up to an event as enormous as the Toba supereruption. As a starting point, the assessment of the possible environmental impacts of Toba can be roughly divided into local, regional, and global effects.

Close to the eruption, all forms of animal and plant life caught up in the collapse of the caldera roof into the depleted magma chamber would have been obliterated. Using a conservative estimate of 15 miles for the radius or size of the YTT caldera translates to an area of about 700 square miles of everything and anything that would be engulfed in the caldera collapse.

PYROCLASTIC FLOWS

Outside the caldera boundary and extending perhaps as far as 60 to 125 miles is the realm of pyroclastic flows. A pyroclastic flow will destroy nearly everything in its path. With rocks ranging in size from ash to boulders traveling along the ground at speeds of 60 miles per hour, these flows knock down, shatter, bury, or carry away nearly all objects in their way. The extreme temperatures of rocks and gas inside the flows, generally between 400 and 1,300°F, can cause any combustible materials, including plants and any life forms, to burn. Even on the margins of pyroclastic flows, death and serious injury to humans and animals may result from burns and inhalation of hot ash and gases.

ASH

In the regional to global distance range, the eruption of Toba formed a very widespread ash fall deposit over the equatorial oceans and southern Asia. A layer of ash with an estimated thickness of 6 inches reportedly accumulated over the entire Indian subcontinent, with similar thicknesses reported over much of Southeast Asia. Even today, at one site in central India, the ash layer is still as much as 20 feet thick. It's possible that this thick layer of ash was the accumulation of wind-blown or water-driven ash. Thicknesses of tens of inches were the norm. It must be realized, however, that ashfalls of tens of centimeters would exterminate most plant and animal life in the effected areas.

Recently, Toba ash has been found in the Arabian Sea and in the South China Sea, the latter implying that several inches also covered southern China. It seems that a new scientific report is published nearly every year announcing the finding of Toba ash further and further from the source region. It's most likely that the ash volume of 190 cubic miles is a minimum estimate and, if revised and added to the tuff volume, would push the Toba eruption unequivocally into second place on the list of largest known explosive eruptions, right behind first-place La Garita Caldera, Colorado, with an unimaginable estimated bulk volume of 1,200 cubic miles.

Toba produced approximately 300 times more volcanic ash than the 1815 Tambora eruption on the Indonesian island of Sumbawa. Given the effect that Tambora had on the regional and global climate, we can only try to imagine what worldwide effects the Toba supereruption would have had. For instance, it has been estimated that as much as 0.8 percent of the material ejected during the Toba eruption could have been in the form of fine dust less than 2 microns in diameter, for a total of about 20,000 Mt of volcanic dust. If only 10 percent of this dust were injected into the stratosphere, conditions of total darkness could have existed over a large area for weeks to months.

In Chapter 9, we tell the story of what happened more than 10 million years ago to a diversity of mammals who were gathering at a watering hole in modern-day northeastern Nebraska when an ash cloud drifted over and the deadly ash started falling on them like "gray snow." The ash was from an explosive volcanic eruption hundreds of miles west in present-day Idaho. The animals died off in a matter of days and weeks by a malady known as Marie's Disease. Death occurs by suffocation when moisture in an animal's lungs combines with the small, jagged pieces of glassy material that constitute the volcanic ash and forms a kind of cement. After they laid down for the last time, their bodies were buried by ash and sandstone deposits, and preserved in place until their fossil skeletons were uncovered 10 million years later in the 1970s. The remarkably well-preserved articulated skeletons are on display at the Ashfall Fossil Beds State Historical Park in Antelope County, Nebraska. There are two lessons to be learned from this story: Ashfall at great distances from a source volcano can have devastating effects on animals, and humans as well, in a very short period of time. The second lesson is that, in this particular case, the source volcano was not a supervolcano but, in fact, was at least an order of magnitude smaller than Toba.

AEROSOL CLOUD

In Chapter 3, we saw that the aerosol cloud from the 1991 Mount Pinatubo eruption circled the globe within several months of the eruption. In addition to spreading throughout the northern hemisphere, the tropical location of Toba, at about 2.5 degrees North, would have most likely resulted in the aerosol cloud spreading into the southern hemisphere within a short time after the eruption. In order to estimate any atmospheric effects that might last for several years or more, it is necessary to calculate a quantity referred to as the optical depth, resulting from the presence of sulfuric acid aerosols in the atmosphere.

Optical depth is a measure of transparency, and is defined as the fraction of radiation, or light, that is scattered or absorbed on a path. One way of visualizing optical depth is to think of a dense fog. The fog between you and an object that is immediately in front of you and can be easily seen has an optical depth of zero. As the object moves away, there is an increasing amount of fog between you and the object, and the optical depth increases proportionally.

Finally, when the object is so far away that it is no longer visible, the optical depth reaches its maximum value.

Based on estimates of the amount of sulfuric acid aerosols injected into the atmosphere by Toba, the globally averaged optical depths range from 1 to 10. The high end value, 10, translates to a reduction in sunlight of about 90 percent, which is typical of an overcast day. Experiments with young grass plants subjected to varying light intensities indicate that photosynthesis is reduced by 85 percent for a decrease in noon-day light by 90 percent. If, in fact, the high end of the optical depth range exceeds a value of 10 as predicted by older data sets, there could be a complete shutdown of photosynthesis for as long as the high optical depth conditions last.

The injection of massive quantities of dust and sulfuric aerosols into the stratosphere from the Toba supereruption resulted in what Rampino and Ambrose, in the Geological Society of America Special Paper 345, referred to as a "volcanic winter." The Toba aerosol cloud is estimated to have caused temperature decreases in the tropics to near or below freezing, hard freezes at midlatitudes, and an extended period of global cooling of 5 to 9°F or more. The cold temperatures in the tropics would have been especially devastating given that most of the vegetation there lacks cold hardiness. Even at higher latitudes, forests and grasslands likely suffered widespread destruction, with recovery periods up to several decades. In addition, studies incorporating the state-of-the-art climate models indicate that the long-lived aerosols, while not of sufficient magnitude to cause an ice age, likely enhanced a long-term (1,000 year) global cool down that was already underway.

EFFECTS ON THE OZONE LAYER

As discussed earlier, one of the potentially most deadly effects of a massive explosive injection of volcanic gases into the stratosphere by a supereruption is the relative depletion of the ozone layer. In the abstract of a paper given at the December 2006 Fall Annual Meeting of the AGU, Ilya Bindeman and others state that results of their study of the oxygen isotope geochemistry of stratospheric sulfate aerosols absorbed on volcanic ash particles imply that massive eruptions are capable of drying out the stratosphere from water, and causing the temporal depletion of the ozone layer. They conclude that the magnitude of the depletion may be many times that of the measured 3 to 8 percent depletion following the 1991 Mount Pinatubo explosive eruption (see Chapter 3), depending on how ozone depletion scales with the amount of sulfur dioxide released into the stratosphere.

If a significant depletion of the ozone layer occurred following the Toba supereruption, the flora and fauna existing at that time could have been hit with a one-two punch. The first punch would have been the climatic impact. Then, as the dust and the aerosols and other volcanic gases started to settle out of the atmosphere, the survivors would be hit by the second punch, the deadly incoming ultraviolet radiation (the UV-B), normally screened by the protective ozone layer.

In their Special Paper 345, Rampino and Ambrose note that the Toba supereruption "occurred during a window of time in which the early human population suffered an extreme bottleneck, with some estimates of as few as 3,000 individuals, followed by the expansion of modern humans." Additionally, "botanical studies of the expected damage to natural ecosystems from a severe cooling and drought such as expected in the aftermath of Toba predict a global environmental disaster that could have contributed to population crashes of various organisms." The extent and ramifications of a bottleneck on the human species is examined in more detail in the following chapter.

SECOND TOBA?

Could an eruption similar to Toba happen in the near future? While the repose period of several hundred thousand years for supereruptions in the caldera complex would suggest that we should not lay awake at night worrying about another Toba, signs of unrest within the caldera point to the fact that this volcanic system is alive and waiting to strike. A warning that the subduction zone that ultimately gives birth to the supereruptions of the Toba complex is alive was given on December 26, 2004. On that day, the devastating Sumatra-Andaman earthquake of magnitude M 9.3 struck along the Sunda arc where the Indo-Australian tectonic plate stick-slips its way under the Eurasian plate, mainly by way of occasional megathrust earthquakes. The December 2004 earthquake was one of the largest earthquakes in more than 100 years, second only to the 1960 Great Chilean Earthquake of M 9.5. Thus, the insatiable subduction factory consumes oceanic crust and continues to disgorge magma to the chamber beneath the Toba caldera complex—to erupt again at some future time.

CHAPTER 6

Bottleneck: A New Evolution

The catastrophic eruption of Toba changed the landscape and altered the climate, but it also changed the course of human evolution, driving the species toward possible extinction. The few survivors, anywhere from 2,000 to 8,000 or more, depending on various estimates from population geneticists and historians, became the mothers and fathers of a new line of human development. According to a stunning theory, each and every one of us alive today is a descendant of the survivors of Toba.

It has long been hotly debated just how, and when, modern humans originated. Two major theories have emerged: one that posits modern humans arose in one place, and the other suggesting premodern humans migrated from Africa to populate the rest of the world. These theories are known as the Multiregional Continuity Model, and the Out of Africa Hypothesis.

TWO COMPETING MODELS

The Multiregional Continuity Model (MCM) claims that after *Homo erectus* left Africa and dispersed into other areas of the Old World, regional populations slowly evolved into modern humans. The MCM posits that:

✳ Some level of gene flow between geographically separated populations prevented speciation after the dispersal.

✳ All living humans derive from the species *Homo erectus* that left Africa nearly 2 million years ago.

✳ Natural selection in regional populations is responsible for the regional variants, sometimes called "races," that we see today.

✳ The emergence of *Homo sapiens* was not restricted to any one area, but occurred throughout the entire geographic range where humans lived.

The "Out of Africa" theory posits, in contrast:

✳ After *Homo erectus* migrated out of Africa, the differing populations became reproductively isolated, evolving independently, and into separate species such as *neanderthalis*.

✳ *Homo sapiens* arose in one place, probably Africa (this geographically includes the Middle East).

✳ *Homo sapiens* later migrated out of Africa and replaced all other existing human populations without interbreeding.

✳ Modern human variation is a relatively recent phenomenon.

Courtesy of "Origins of Modern Humans: Multiregional or Out of Africa" by Donald Johanson for ActionBioscience.org.

OUT OF AFRICA

Recent genetic and geological evidence is swaying more minds in favor of the Out of Africa model. In a May-2001 article titled "Boost for 'Out of Africa' Theory" for the BBC News, reporter Ivan Noble points to the study by Li Jin of Shanghai's Fudan University and Spencer Wells of Wellcome Trust Centre for Human Genetics in Oxford. This study examined the

Y chromosomes of more than 12,000 people across Asia, finding no traces of any ancient non-African influence. The study findings, published in the journal *Science*, stated, "This result indicates that modern humans of African origin completely replaced earlier populations in East Asia," and fully supports the Out of Africa model, while "somewhat putting the nail in the coffin of multiregionalism," according to Wells.

While this study used only Y chromosomes, which are passed down through the male line only, the researchers hope to duplicate the efforts with mitochondrial DNA, which is passed down the female line to add to the evidence.

So, if modern humans did indeed emerge out of Africa, what, then, did Toba have to do with their evolutionary history?

TOBA AND EVOLUTIONARY HISTORY

During the time of Toba's supereruption, Europe and Africa were in the Middle Paleolithic period of the early Stone Age, which began approximately 120,000 years ago, and ended 40,000 years ago. This was a time of great change for the human species. Neanderthal man (*Homo neanderthalensis*), a species of the *Homo* genus, inhabited Europe and Western Asia, as well as parts of Northern Africa. Neanderthal remains show them to be skilled hunters, versed in the use of fire and stone tools of the flake tradition, as well as bone needles used for crude forms of sewing animal skins for clothing. They hunted prehistoric mammals for food, and may have painted their dead as part of a religious belief system, though scant evidence is available to support this theory.

Known for their sloping foreheads and prominent brow, Neanderthals were shorter, stockier versions of the emerging *Homo sapiens*. *Homo sapiens* were present in south and east Africa, increasing population and eventually overtaking their Neanderthal counterparts. Some historians and paleontologists believe Neanderthals persisted until about 50,000 years ago, when they vanished from the Asian continent, reaching extinction in Europe some 20 to 30 thousand years later. Also present on the Asian continent was *Homo erectus*, which includes Java Man and Peking Man.

But by 30,000 years ago, this diverse group of hominids vanished and humans everywhere evolved into the anatomically modern man of today. *Homo sapiens* eventually migrated out of Africa, replacing all other human populations, and did this without interbreeding with other hominid species.

Contrary to popular belief, Neanderthals were not sub-species of *Homo sapiens*, but were a separate species from modern man. Some scientists argue that the two species may have interbred while they both walked the Earth, but in an article for ABC News in 2005, Cambridge Professor Paul Mellars stated that there was no evidence for this kind of "cultural interaction." In July of 2006, the Max Planck Institute for Evolutionary Anthropology and the 454 Life Sciences announced they would sequence the Neanderthal genome, to be completed by 2008. Roughly the same size as the human genome, the Neanderthal genome may share many identical genes and could unlock some of the mysteries of human evolution. But for now, genetic testing with both mitochondrial DNA, which we will discuss later in the chapter, and with nuclear DNA, show that Neanderthal DNA is quite distinct from our own, and that no evidence of hybridization or interbreeding has been proven on a genetic basis, despite the two species having co-existed when *Homo sapiens* migrated into the European region.

Just prior to Toba's eruption, emerging man was experimenting with rock art and mastering the use of fire and tools. And man was expanding his food sources, exploiting shellfish, and possibly smoking and drying meat as a means of food preservation. Little could these hearty survivors, used to living with constant danger and the pressure of survival, realize how a nearby eruption could alter their very existence, if not end it altogether.

Those living close to Toba died instantly when the mighty beast erupted, and so did most of the animal and plant life nearby. The survivors, those far enough away on the African and Asian continents, hidden in caves or in isolated tropical pockets, watched as their own numbers diminished due to the climate change and subsequent ice age that Toba ushered in. Without food, fresh water, prey, and shelter from the cold, the population went through a bottleneck, drastically reducing the number of humans from an estimated 100,000 to somewhere between 2,000 to 8,000 or more.

The question then becomes, what happened to the survivors?

Most paleoanthropologists believe that sometime between the last 60,000 and 100,000 years, our human ancestors emerged out of Africa, and eventually replaced the archaic east Asian and Neanderthal populations. Many theories attempt to explain this "Out of Africa" concept of modern human origin, including the Replacement Hypothesis, Weak Garden of Eden Hypothesis, and Multiregional Hypothesis.

OUT OF AFRICA

The origin of modern humans out of Africa focuses on several theories that explain the dispersal and major subdivisions of our species throughout the last 100,000 years. In "Genetic Evidence on Modern Human Origins," Alan R. Rogers and Lynn B. Jorde of the University of Utah describe the competing theories:

✳ **Multiregional Hypothesis:** The major subdivisions of the human species evolved in situ over a long period of time, with gene flow accounting for much of the similarity now observed among groups.

✳ **Replacement Hypothesis:** Earlier populations were replaced some 30,000 to 100,000 years ago by populations of anatomically modern humans that originated in Africa.

✳ **Weak Garden of Eden Hypothesis:** A small ancestral human population separated into several partially isolated groups approximately 100,000 years ago, then, some 30,000 years later (around time of Toba) these groups underwent either simultaneous bottlenecks or simultaneous expansions in size.

Using evidence gleaned from fossils and geological studies and genetic research, these three theories have been joined by a more recent, and more intriguing, alternative theory, proposed by Professor Stanley H. Ambrose of the Department of Anthropology at the University of Illinois.

Known as the Toba Catastrophe Theory, Ambrose proposed the idea in a 1998 paper published in the *Journal of Human Evolution* titled "Late Pleistocene Human Population Bottlenecks, Volcanic Winter, and Differentiation of Modern Humans." The theory suggests that the impact of the eruption of Toba approximately 70,000 to 75,000 years ago led to a volcanic winter and a population bottleneck that drastically reduced the size of the human population. Ambrose suggests in his article that his theory could finally resolve a central question of a problematic paradox of the African origin theory of humanity: That paradox asks, "If we are all so recently out of Africa, than why do we not all look more African?"

In a September 8, 1998 article for *Science Daily* titled "Ancient Volcanic Winter Tied to Rapid Genetic Divergence in Humans," Ambrose stated, "The standard view of human evolution has been that modern populations evolved from an ancient African ancestor. We assumed that they differentiated gradually because we assumed ancestral populations were large and stable." But new genetic research instead points again and again to dramatic changes in population size over the course of human history.

The population bottleneck created by Toba, and its aftereffects, could be the answer to this question of why modern human races differentiated abruptly, right around the time of the eruption, instead of over a gradual period of more than a million years.

Ambrose also challenged the Multiple Dispersals model of the Out of Africa theory, which stated that the first dispersal of anatomically modern humans took place approximately 100,000 years ago, based upon early modern human skeletal remains found in the near East. But Ambrose felt this first dispersal did not succeed at permanently establishing modern humans out of Africa. He was backed up by genetic evidence that pointed to non-African populations dividing into two distinct populations approximately 50 to 75,000 years ago.

To understand the theory Ambrose proposes, and examine the genetic evidence backing his theory, we first need to understand what a population bottleneck is, why they happen, and how they change the course of evolution for a particular species.

BOTTLENECKS

To state it simply, a bottleneck occurs anytime a population experiences a drastic reduction in size for the extent of at least one generation. Bottlenecks can be the result of global climate changes, asteroid impacts, volcanism, or any other catastrophic event with massive environmental repercussions that result in the decimation of a species or a genetic line. Bottlenecks also reduce a specific population's genetic variation enough to prevent that population from adapting to its newly created environment. Bottlenecks also increase both genetic drift, and the rate of inbreeding because of the limited pool of potential mates.

Population bottlenecks have occurred in the animal kingdom many times in the past. In the early 20th century, the European bison, or wisent, underwent a bottleneck that dwindled their population to approximately 360,000, all of which can be traced to 12 individual bison. Even the American Bison experienced such a bottleneck due to over-hunting, which saw their numbers reach extremely low levels before rebounding, as many species do after recovering from a bottleneck.

Northern elephant seals saw their population reduced to about 20 individuals around the year 1890, thanks again to over-hunting by humans. Even though the species has made a comeback, now numbering approximately 127,000, their genetic variation is still much less than other species of southern elephant seals lucky enough to have not been hunted to near extinction. Their

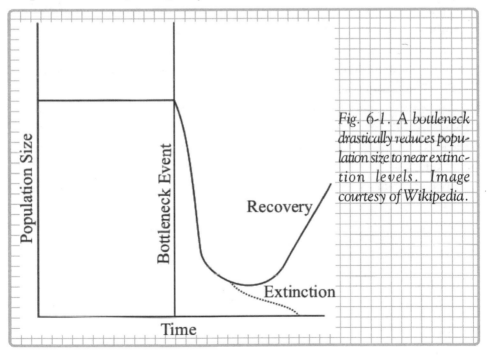

Fig. 6-1. A bottleneck drastically reduces population size to near extinction levels. Image courtesy of Wikipedia.

genetic similarity means, in some ways, that all surviving Northern elephant seals are brothers and sisters, and the lack of genetic variation makes the species vulnerable to new diseases.

In a paper titled "Genetic Diversity and Conservation of Endangered Animals Species" published in *Pure and Applied Chemistry* issue 74, 2002, a group of scientists examined the loss of biodiversity that results from extinctions. Their paper focused on the giant panda population, from which they genetically sequenced 32 individuals from four regional populations. They wanted to examine the link between a species' population history and genetic structure, and the development of more sound conservation plans for endangered species.

The scientists found that the giant panda, found only in China, may have indeed undergone a population bottleneck approximately 43,000 years ago. They also discovered genetic evidence, proving that a species of primate called the snub-nosed monkey also underwent a bottleneck around the same time, giving further credence to their bottleneck theory for the panda (if one species suffers, then surely evidence of others should appear as well). The low genetic variation found in both species indicates a bottleneck, but, as the authors of the paper point out, also suggests possible inbreeding and an increased vulnerability to environmental changes. Species that have undergone a bottleneck get a double whammy, in that they lack the genetic "strength" to fight off pending extinction if the bottleneck's cause, in this case human hunting and destruction, continues unabated.

The smaller and more isolated a species population is during and after a bottleneck, the more vulnerable it is to extinction. But it doesn't always mean the end of a species outright. The golden hamster underwent such a drastic bottleneck—most of the remaining individuals of the species can be genetically traced back to a single litter from the Syrian desert in the 1930s. Cheetahs are thought to be the victims of an extreme bottleneck as well, which would explain why they are so closely related to one another; so much so that a skin graft from one cheetah to another cheetah fails to provoke an immune response—an indicator of bottleneck presence. Both these species continue to exist, although that existence is fragile, mainly because of the human threats of hunting, poaching, and environmental degradation.

While humans have certainly been the culprits behind many animal species bottlenecks, they have also been the victims as well. Paleontologists, anthropologists, and geneticists alike have often wondered why human beings exhibit so little genetic variation. On the surface, we all look quite diverse, yet our genes do not show that same diversity. The bottleneck that occurred at the time of Toba could be the reason why; it could also explain why differentiation among humans is more a recent phenomenon than one from our distant past.

In a May 2006 article for *BBC News*, Dr. David Whitehouse wrote about the near-extinction of humans about 70,000 years ago, stating that this date seems to also be aligned with the journey of *Homo sapiens sapiens*, the anatomically correct modern humans, out of Africa. He comments on the mystery of why a small group of our closest genetic relatives, the chimps, have more genetic diversity than all of the billions of humans alive today combined (all humans have DNA that is virtually identical).

This suggests that around that same time of migration, something happened to reduce the gene pool small enough to wipe out genetic variation. Whitehouse points to research by geneticists from Stanford University, and the Russian Academy of Sciences that compared 377 microsatellite (short repetitive DNA sections whose patterns of variation differ among populations) markers in DNA from 52 various regions worldwide, and earlier studies that both pointed to the first human migration out of Africa occurring sometime between 66,000 and 100,000 years ago. The size of this migration could have been as low as 2,000 humans, and, as Whitehouse stated, "It was out of this small population, with its consequent limited genetic diversity, that today's humans descended."

This May 2003 study, which was published in the *American Journal of Human Genetics*, suggests that the genetic findings are consistent with the "Out of Africa" theory, and point to the fact that because microsatellites are passed from generation to generation, they are useful for estimating when two populations may have diverged.

DNA AND HUMAN EVOLUTION

Most scientists now agree that sub-Saharan Africa seems to be the cradle of human evolution, though many disagree as to how many times humans migrated out of Africa, and when; but most agree on the fact that these migrations seem to be relatively recent. Geneticists also agree that all living human populations are traceable along maternal lines to an African female living approximately 120,000 to 200,000 years ago. Known as the "Eve Hypothesis," this theory claims that, using mitochondrial DNA (mtDNA), which is inherited only from one's mother, all humans have a common female ancestor. If we look at genetic history as a tree, all branches or sequences coalesce into one: "Mitochondrial Eve," the African female thought to be the most recent common matrilineal ancestor of all humans on Earth today.

To be fair, there is a Y chromosome Adam, but he has been dated much more recently, somewhere between 60,000 and 90,000 years ago. The Y chromosome Adam concerns patrilineal descent, but does not leave a similar trace in women today as does Mitochondrial Eve. Mitochondrial Eve and Y Chromosome Adam are considered "most recent common ancestors," or MRCA, yet some geneticists believe they are not necessarily the MRCAs of those of us alive today. However, most do agree that our MRCA did live during the Paleolithic era.

MITOCHONDRIA

Mitochondria are the energy-generating structure that is just outside the nucleus of a cell. Because mitochondria are present in such large numbers in each cell, fewer samples are necessary for DNA tests, and because mtDNA mutates at a steady rate, geneticists use it to determine a sort of "historical molecular clock" that allows them to trace the evolution of a species. Scientists have been looking to mtDNA as a means of understanding our genetic history since the early 1990s. The nuclei of our cells store most of our genetic information, but a small and separate quantity of genetic information is stored in the mitochondria. In fact, Mitochondria have their own genes, and this DNA is distinct from the rest of the cell's DNA. Mitochondria also have a higher rate of substitution—mutations where a nucleotide is replaced with another nucleotide. They also do not recombine and mix up sections of DNA from the mother and father, which can create a muddied genetic history.

Mitochondrial DNA is also important because any mutations are passed on from mother to child through generations at a regulated rate. Over a period of time, these mutations accumulate in the descendant population. Scientists can look at the rate of mutation in mtDNA to determine the distribution of these mutations and estimate the past size of a given population. This allows the scientists to see patterns in population growth or decline.

WEAK GOE

Using the average rates of genetic mutation one normally sees in a species through time, some scientists now claim that the Toba event could have indeed spawned a new theory of human evolution, a tweaking of the earlier Weak Garden of Eden (Weak GOE) model.

The Weak GOE model suggests that modern humans originated in Africa about 130,000 years ago, and points to the invention of advanced stone tool technology as the means of population growth some 40,000 to 50,000 years ago. But anthropologist Stanley Ambrose of the University of Illinois believes that it was the volcanic winter of Toba, which lasted years after the supereruption and created a population bottleneck, that was really responsible for the expansion of the population after suffering drastic reduction in size, rather than the emergence of tool technology.

Ambrose began to formulate his theory after he invited Professor of Anthropology Henry Harpending of the University of Utah to come and talk to his students. Harpending, along with Lynn Jorde, Professor of Genetics at the University of Utah, had studied patterns of mitochondrial DNA within the human population and came to the stunning discovery that something happened between 70,000 and 80,000 years ago, something so big it literally reduced the human population down to as few as 5,000 individuals.

Ambrose listened attentively as Harpending talked about this bottleneck. As the lecture continued, Ambrose broke out in a sweat as he wondered what might have caused this amazing bottleneck. Excitedly, he went up to Harpending after the lecture and told him about a paper he had just finished reading that discussed the supereruption of the Toba volcano in Sumatra. That paper was "Climate-Volcanism Feedback and the Toba Eruption of ~74,000 Years Ago," written by Michael Rampino and Stephen Self for the 1993 *Quaternary Research Volume 40*. Harpending didn't jump on the paper right way, but in the BBC program *Supervolcanoes*, he is quoted as saying "I didn't read it until a week later, and when I read it you know it was like somebody kicking you in the face. There it was."

The work of Rampino and Self made sense, and was the final puzzle piece that completed the bigger picture. The eruption of Toba happened right smack in the middle of the time period when the bottleneck occurred.

Ambrose called his exciting new theory the "Weak Garden of Eden/ Volcanic Winter" model, but it has become more widely known in recent years as the Toba Catastrophe Theory. The theory has gained quite a bit of favor recently, with additional research confirming Ambrose's idea that the six-year climate change caused both the bottleneck, and the emergence out of Africa shortly afterward.

The Toba Catastrophe Theory, or TCT, blames the massive environmental change created by Toba's eruption, which decreased global temperatures by 5 to 9°F or more for approximately six years, for the population bottlenecks that eventually led to the extinction of all human species except for the one from which we all descended. The rapid amount of differentiation seen in humans at this time is directly related, according to the TCT, to the low population levels that are needed to accelerate evolutionary change.

In an interview for *Science Daily* in 1998, Ambrose stated, "The standard view of human evolution has been that modern populations evolved from an ancient African ancestor. We assumed that they differentiated gradually because we assumed ancestral populations were large and stable." Ambrose refers to the research of volcanologists such as Michael Rampino and Steven Self, which corroborates his theory that the brutal volcanic winter of the Toba event created a "global human population crash." Ambrose went on to state, "When our African ancestors passed through the prism of Toba's volcanic winter, a rainbow of differences appeared."

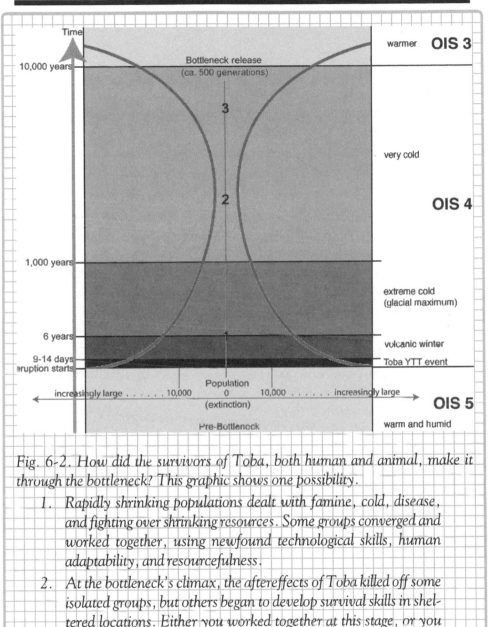

Fig. 6-2. How did the survivors of Toba, both human and animal, make it through the bottleneck? This graphic shows one possibility.

1. Rapidly shrinking populations dealt with famine, cold, disease, and fighting over shrinking resources. Some groups converged and worked together, using newfound technological skills, human adaptability, and resourcefulness.

2. At the bottleneck's climax, the aftereffects of Toba killed off some isolated groups, but others began to develop survival skills in sheltered locations. Either you worked together at this stage, or you perished.

3. Surviving populations adapted and began the road to recovery, increasing their population, skills, and tool technology.

OIS5 refers to Oxygen Isotope Stage 5: hot and humid (130,000 to 73,000 years ago.

OIS4 refers to Oxygen Isotope Stage 4: cold and dry (73,000 to 63,000 years ago, coldest OIS for past 110,000 years).

OIS3 refers to Oxygen Isotope Stage 3: warm and humid, but milder than OIS5 (63,000 to 45,000 years ago).

Diagram and descriptive text courtesy of Toba, Volcano by George Weber of the Andaman Association.

After Toba, humans once again began to populate and fan out of Africa to places such as Indochina, Australia, the Middle East, Asia, and beyond, with dramatic growth occurring approximately 50,000 to 60,000 years ago in genetically isolated populations. This coincides with Ambrose's belief that the duration of the Toba bottleneck could have been about 10,000 years, with "release" about 60,000 years ago. During the time of release, the population would increase and dispersal would become evident, along with increased human differentiation as the populace grew and bred.

Interestingly, genetic evidence shows a division approximately 50 t o 75,000 years ago of non-African populations into two groups: southern Australasian and northern Eurasian.

TOBA CATASTROPHE THEORY

In his paper titled "Late Pleistocene Human Population Bottlenecks, Volcanic Winter, and Differentiation of Modern Humans," Ambrose sets the foundation of his Toba Catastrophe Theory.

The volcanic winter into which Toba plunged the world slashed the human population enough to create the stage for rapid changes in the surviving humans. The six-year period after the eruption deposited the largest amount of volcanic sulphur onto the planet in the past 110,000 years. The previous chapter discussed the catastrophic changes brought about by Toba, and Ambrose makes the comparison between the "year without a summer" that Mt. Tamboura created as described earlier, and the six years without a summer that Toba unleashed on the Earth.

Ambrose's theory also states that the bottleneck may have lasted only 1,000 years, during only the coldest part of the volcanic winter that Toba unleashed. But either way, the vast majority of Toba survivors were to be found in the African tropical region, thus resulting in the greatest genetic diversity among survivors in Africa.

Ambrose's TCT suggests that the Toba event is the reason why human mitochondrial DNA and Y Chromosome DNA show coalescence around the same time the supereruption occurred approximately 74,000 years ago. All humans' female and male lines of ancestry seem to be traceable to a relatively small group of individuals alive at the time. His theory also explains the DNA mutation rate, which increased temporarily after Toba, as evidence of a bottleneck during the last glacial period.

Coalescence does not automatically indicate a bottleneck, and some scientists believe a bottleneck would only be indicated if the female line and male line ancestries coalesced more recently than expected on a genetic timetable. Other scientists support the "long bottleneck theory," suggesting that a transplanting model or a longer duration population bottleneck and recovery period is responsible for the limited genetic variation in humans.

INCREASING EVIDENCE

But others disagree, pointing to increasing evidence is coming out of a variety of laboratories working to turn back the genetic clock—evidence linking the Toba event to a more recent, distinct bottleneck that left its mark on all of us alive today.

Lynn Jorde, in the BBC program _Supervolcanoes_, stated that "Our population may have been in such a precarious position that only a few thousand of us may have been alive on the whole face of the Earth at one point in time, that we almost went extinct, that some event was so catastrophic as to nearly cause our species to cease to exist completely." But the few that did survive left us a genetic record of their regrowth and dispersal out of Africa and onto other continents, where the human species thrived and prospered. The work of geneticists, anthropologists, and scientists converged to paint a picture of survival that also filled in the long-missing pieces of a puzzle. That puzzle, the dramatic reduction of genetic diversity, paralleled the dramatic reduction of the human species that occurred when Toba erupted, and breathed new life into our understanding of why we, as a species, share such a lack of diversity at the DNA level, despite our numerous physical differences.

In other words, we look different, but at the most minute level of our being, we are all the same, all from the same "family tree."

Further research by Lynn Jorde and Alan R. Rogers showed a similar population decline and rebirth for another species living alongside modern humans at the time of the supereruption. In a paper titled "Genetic Evidence on Modern Human Origins," published in the February 1995 edition of the journal, *Human Biology*, the two researchers examined "mismatch and intermatch distributions." Mismatch distributions from a subdivided population use pairs drawn from a single subdivision. Intermatch distributions use pairs drawn from different subdivisions. The patterns that emerge create a "wave" that reflects two possible types of events: First, the event may be an expansion that occurred prior to the separation of the ancestral population. The second, according to the paper, occurs when there is no wave in the original distribution. By examining wave patterns in three pairs of populations, the researchers suggest that the wave in the human data that points to a Toba-related bottleneck should also be evident in other species.

Jorde and Rogers found evidence of similar mismatch distribution waves of humans and the Eastern chimpanzee. Though they concede that the similar waves could be the result of the "advance of a favorable mutation," it seems unlikely that this could happen to both species at the same time. They conclude, "Thus, this evidence supports the hypothesis that the mismatch waves reflect changes in population size rather than natural selection. It also suggests that the waves of both species reflect some environmental catastrophe."

The paper goes on to explore the comparisons between chimpanzees, which display mitochondrial diversity in excess of humans, and our own strangely lacking diversity, despite our massive population. The researchers point to a state of equilibrium in which genetic statistics converge geometrically at just the right rate. If the statistics converge too slowly, we can learn little about the "demographic perturbations" that may have occurred to a given species. If the statistics converge too quickly, we are unable to reach far back into the past enough to get usable data. Thus, rates of convergence are critical to reaching a state of genetic equilibrium that will allow us a glimpse of our own past.

At equilibrium, the diversity of selectively neutral genes should be low in small populations, and high in larger ones," the paper states. Molecular assays, an analytical procedure to test the properties and/or composition of organic substances, metals, drugs, or other materials by chemical means, confirm an unusual discrepancy, according to Jorde and Rogers. These essays show that human mitochondrial diversity is lower than many more numerous species, yet is also lower than that of anthropoid apes, which have a smaller population size. The researchers conclude that our human population expanded from a small beginning in the past 150,000 years, and that there has not been enough time for the species to reach that state of equilibrium. This points, again, to evidence of an event in our historical past "such as a bottleneck in population size, rather than a long history of moderate population size.

This paper concludes that "all available genetic evidence is consistent with the proposition that the major human populations separated from a small initial population roughly 100,000 years ago, and that most of these separate populations experienced a bottleneck, or an episode of growth, several tens of thousands of years later."

Henry Harpending, along with fellow researchers Mark Batzer, Michael Gurven, Lynn Jorde, Alan Rogers, and Stephen Sherry, wrote a paper for the February 17, 1998 special series of Inaugural Articles by special members of the National Academy of Sciences. The paper, titled "Genetic Traces of Ancient Demography," covered a number of different genetic techniques for determining patterns of gene differences among humans, and what information they contain about our demographic past.

In summary, the estimates of overall human effective size of 10,000 from nuclear sequences, Alu insertions, and HLA exons, mtDNA mismatch distributions, frequency spectra from mtDNA, and from the Y chromosome, and discordance between allele size variance and homozygosity at tandem repeat loci all support the hypothesis of a bottleneck in our past during which the number of our ancestors was only a few thousand breeding adults...." The team of researchers, from which Stanley Ambrose then formulated his TCT, went on to state, "The best available estimates of mtDNA mutation rates imply that the expansion occurred between 100,000 and 50,000 years ago in excellent agreement with archeological

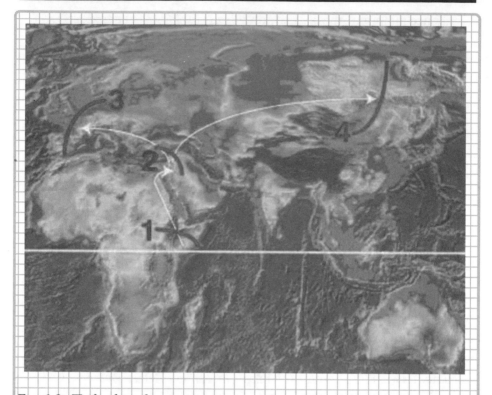

Fig. 6-3. Technological invention came out of post-Toba Africa beginning approximately 50,000 years ago. In Africa, the main periods of stone tool technology carry the following terminology: ESA (Early Stone Age); MSA (Middle Stone Age); LSA (Late Stone Age). In Europe and Asia, similar but later technologies are known as: LP (Lower Paleolithic); MP (Middle Paleolithic); UP (Upper Paleolithic). This map shows:

1. The transformation of MSA to LSA in East Africa, approximately 50,000 years ago.
2. The transformation of MP to UP in the Levant, approximately 47,000 to 43,000 years ago.
3. The transformation of MP to UP in Western Europe approximately 43,000 to 40,000 years ago.
4. The transformation of MP to UP in Siberia approximately 43,000 to 40,000 years ago.

Image and map description courtesy of Toba, Volcano by George Weber, Andaman Association.

evidence of the earliest modern humans about 100,000 years ago, and the 'creative explosion' of upper Paleolithic type technology about 50,000 years ago...."

But because mtDNA and Y chromosomes coalesced several hundred thousand years ago, it makes it difficult for the researchers to pinpoint an exact population size before said bottleneck occurs. They do conclude that the genetic evidence prevails for the theory that we are all descended "from a population that was effectively a separate species for at least the last 1 to 2 million years. Although the size of this population must have fluctuated over time, it was often reduced to the level of several thousands of adults."

It is also difficult to pinpoint the exact size of the surviving population of *Homo sapiens* that could account for genetic uniformity we see today. In his book *Toba Volcano*, George Weber, president of the Andaman Association, refers to an estimate by Harpending et. al. of 40 to 600 females that came through the bottleneck. This would translate into under 3,000 individuals. Other estimates by other researchers suggest 500 to 3,000 females (A.R. Rogers, 1993) and even 1,000 to 4,300 individuals (F.J. Ayala, 1996). Ambrose himself suggests that number may have been even higher, with approximately 10,000 reproductive-age females. No matter which estimate comes closest, Weber concludes that in the not-too-distant past, our human population dwindled to the size of "a small country town."

Weber speculates that the surviving population lived in a relatively small area, most likely East Africa. "The hardest hit area then inhabited by *Homo sapiens* was the Middle East, including probably the Arabian Peninsula." He goes on to state, "if there were modern humans in India at the critical moment, they would have had no chance to survive at all...it is likely that *Homo sapiens* groups living in Asia were wiped out (along with their genetic diversity)." Referring to this region as the "Toba kill zone," Weber states that the close proximity of the most intense ash fall would have wiped out all life, human or otherwise.

The further south the population lived at the time of the supereruption, the more likely they were to survive. Weber, along with many others who support the Toba Catastrophe Theory, believes that the "most likely place for human survivors is along the eastern coast of Africa, possibly on the southern side of the equator. Of all the areas known to have had human or human-like inhabitants at the time of Toba, this was the least affected."

Interestingly, Weber goes on to examine why *Homo sapiens* were hit so hard by the bottleneck, at least in comparison to the lesser affected primate species that also suffered their own mild bottlenecks. He suggests that African apes of today all live on the western side of the extensive mountain chains that run north-south along the east coast of the continent. Because their population was, at the time of Toba, already dispersed in this area, only the apes east of the mountain did not survive.

Homo sapiens, in contrast, "had left most traces of its development and existence along the eastern coast of Africa where the species had spread on a mostly vertical north-south axis along the mountains and the coast, especially in Kenya and Ethiopia." *Homo sapiens* expanded out of Africa into the rich and fertile valley of the Nile to the Mediterranean African coast, Weber points out, and then into Western Asia. "Unhappily, the geographical distribution of *Homo sapiens* 73,000 years ago ensured the almost maximum exposure possible to the direct and indirect effects of Toba."

Similar to the great apes that may have taken shelter in mountain caves, the humans that did survive must have found a way to find shelter from the falling ash, the climate changes, and the harsh volcanic winter that would last for approximately six years. Weber believes three things were tantamount to who, or what, survived:

1. Because of the rapidly shrinking population, famine, cold, disease, and lack of water, neighboring groups either worked together or fought for resources, with necessity forcing adaptive cooperative skills, new technical capabilities, and "a remarkable degree of adaptability and resourcefulness."

2. Aftershock may have killed off isolated groups, or groups died off that were unwilling to cooperate or adapt fast enough.

3. The surviving humans began to recover, reproducing and learning new skills and technology.

Just as the genetic evidence supports our connectedness to every Toba survivor, the very means of survival call for understanding that connection, cooperation, and adaptation are much more conducive to making it out of a major catastrophe alive than sheer force of will or brute strength. An embracing of technology also helps. In the May 28, 2003 *Science Daily* article titled "Scientists Use DNA Fragments to Trace the Migration of Modern Humans,"

Stanford University geneticist Marcus Feldman extrapolates on this when he compared the lack of population growth over time of indigenous hunter-gatherer societies in Africa, Oceania, and America to the higher rebound rates of the ancestors of sub-Saharan African farming populations that experienced an expansion around 35,000 years ago. "This increase in population sizes might have been preceded by technological innovations that led to an increase in survival and then an increase in the overall birth rate." The populations of Eurasia and Asia also show a similar marked expansion around 25,000 years ago.

Despite past belief that only two human species came through the bottleneck—*Homo sapiens* and the stocky and cold-hearty *Neanderthalis*, who survived Toba only to meet their extinction in the Middle Eastern region around 30,000 years ago—Weber and other scientists agree that two other species survived, and possibly others yet to be discovered. One of those is *Homo erectus*, once thought to be extinct more than 300,000 years ago. Remains of *Homo erectus* found in the central Java area of Indonesia were dated at approximately 30,000 years ago, making *Homo erectus* a contemporary of modern man. The other surviving species is the intriguing miniaturized *Homo florensiensis*, the tiniest and most recently discovered member of the genus *Homo*, believed to have been alive and well up until about 12,000 years ago, the date of remains found on the Flores Island in Indonesia. Yet ultimately, of all these species, only *Homo sapiens* did not meet with complete extinction.

That *Homo sapiens* are the only surviving species of human today, despite having very little genetic diversity to show for our long evolutionary history, continues to intrigue anthropologists, geneticists, and volcanologists alike, all of whom hold a piece to the puzzle of our past. According to Professor Mike Archer, director of the Australian Museum, in an interview for ABC NewsOnline (June 11, 2003), "We should have had much more variation than we've got."

But despite our genetic eccentricity, *Homo sapiens* is now the most populous of any large species on Earth, and seemingly can live, and even thrive, in any climate or locale. We survived The Big One once before.

The question remains: Can we survive the next Big One? As Professor Mike Archer told ABC NewsOnline, "It does highlight the vulnerability of being human....We can't presume that because we've been here for, you know, at least 70,000, some would argue as much as 300,000 years as a species, that we're going to go on forever."

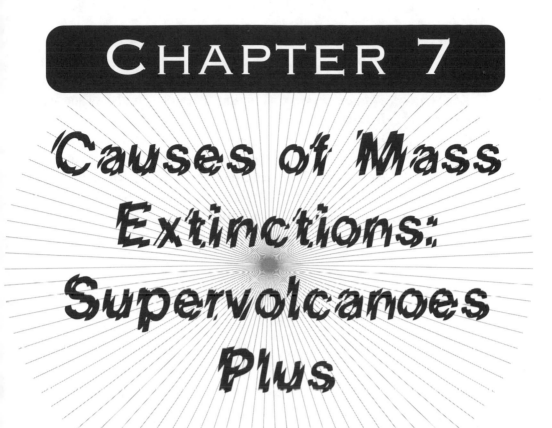

CHAPTER 7

Causes of Mass Extinctions: Supervolcanoes Plus

In the previous chapter, we saw how humans underwent a genetic bottleneck thought to have been caused by the eruption of the Toba supervolcano approximately 74,000 years ago. In this chapter, we will focus on the natural causes of extreme forms of bottlenecks that have occurred episodically during the past several hundred million years. These extreme bottlenecks are known as mass extinctions, and are periods when multiple species of animal and plant life ceased to exist. Both terrestrial and extraterrestrial causes will be considered, in particular the role of supervolcanoes.

Paleontologists study the forms of life that existed in prehistoric or geologic times as represented in the fossil record of animals, plants, and other organisms, and have defined a background rate of extinction as the number of extinctions that would be occurring naturally in the absence of human influence. They estimate this rate to be one to 10 species per year for the past several hundred million years. For every species that is alive today, an estimated 1,000 species have lived at some time in the past and become extinct.

In other words, of all the species that have ever existed, about 99.9 percent are now extinct. While most of these species died off before humans appeared on the scene, the existence of a fraction of the total number of species is found in the fossil record.

Occasionally, the background rate is significantly exceeded, and episodes of heightened extinction are recognized in the fossil record. Estimates for the number of significant extinctions during the past several hundred million years vary from as few as five to more than 20. The current consensus among paleontologists is that, within the past 450 million years, there were five major mass extinctions or episodes when unusually large numbers of species became extinct. While numerous investigations completed in the past 20 to 25 years indicate that these "Big Five" mass extinctions occurred as a result of major changes in the prevailing global ecological conditions, a controversy over what caused the changes in the global ecological conditions continues to the present day.

The discussion in the following sections of this chapter refers to divisions of the geologic column that geologists have pieced together from layers of fossil-rich rocks and sediments from different parts of the globe. In order to facilitate the discussion, the principal divisions of the geologic column are outlined in the following: The divisions of main interest, in order of decreasing time intervals, consist of eras, periods, and epochs. In the following, *mya* stands for "million years ago." There are four eras:

✳ Precambrian (4,600 to 545 mya)
✳ Paleozoic (545 to 248 mya)
✳ Mesozoic (248 to 65 mya)
✳ Cenozoic (65 mya to the present)

Each era is subdivided into periods. The most important periods for the discussion at hand are the periods of the Paleozoic, the Mesozoic, and the Cenozoic eras.

Within the Paleozoic, the periods of interest are:

✳ Ordovician (490 to 443 mya)
✳ Devonian (443 to 354 mya)
✳ Carboniferous (354 to 290 mya)
✳ Permian (290 to 248 mya)

Periods within the Mesozoic are:

✻ Triassic (248 to 206 mya)

✻ Jurassic (206 to 144 mya)

✻ Cretaceous (144 to 65 mya)

Periods within the Cenozoic are:

✻ Tertiary (65 to 1.8 mya)

✻ Quaternary (1.8 mya to present)

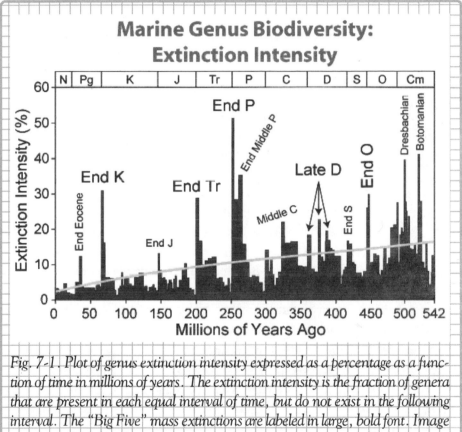

Fig. 7-1. Plot of genus extinction intensity expressed as a percentage as a function of time in millions of years. The extinction intensity is the fraction of genera that are present in each equal interval of time, but do not exist in the following interval. The "Big Five" mass extinctions are labeled in large, bold font. Image courtesy of Wikipedia.

Figure 7-1 is a plot of the "genus extinction intensity," that is, the fraction of genera that are present in each interval of time, but do not exist in the following interval. The genus is used in this plot because it represents a more complete grouping of life forms than the lower category of species. The data are not intended to represent all possible genera that ever lived, rather a selection of marine genera most likely preserved as fossils. The line drawn across the figure curving down to the left is a fit to the data to show the long-term trend. Note that extinction data for the most recent 10,000 years are not included on the plot.

THE BIG FIVE

The ("Big Five") mass extinctions are labeled in large, bold font in figure 7.1 on page 153. Before examining the figure in more detail, there are certain characteristics that the Big Five have in common: They all represent the extinction of a significant component of global faunas and/or floras, and they are all relatively sudden in geological terms, occurring through a few million years at most, and in some cases much shorter time periods than that.

The extinction labeled *End O* occurred approximately between 440 and 450 mya in the Ordovician period, and is considered by many paleontologists to have been the second most devastating extinction to marine communities in the history of the Earth. The extinction labeled *Late D* refers to an event in the Late Devonian, and is broken down in the plot in Figure 7-1 into three distinct events. Other extinction events include the end Permian (*End P*), which is also known as the "Mother of Mass Extinctions," or the "Great Dying," and is considered Earth's worst mass extinction event with approximately 96 percent of all marine species and 70 percent of land species becoming extinct. We will return to the end Permian extinction after consideration of the most recent of the Big Five mass extinctions: the *End K* event at the Cretaceous-Tertiary (KT) transition approximately 65 mya.

One question that arises at this point is: How does the KT designation follow from the Cretaceous-Tertiary period designations? The answer is that the "K" in KT is from the Greek word *Kreta*, or chalk. The "Cret" in Cretaceous also comes from the same word, and points out that there was a lot of chalk laid down during this period.

The *End K* or KT extinction is probably the most famous of the Big Five extinctions. Its fame derives from the fact that this event heralded the demise of the dinosaurs, and, the emergence of mammals as the dominant life form on land. While the demise of the dinosaurs at the end of the Cretaceous was accepted by most geologists and paleontologists, the cause of their demise became a hotly debated issue following the publication of one of the most earthshaking and influential scientific papers in the history of the Earth sciences.

KT EXTINCTION

In 1980, a team of four scientists from the University of California at Berkeley proposed that it was the impact of a meteorite, or an asteroid, about 6 miles in diameter with the Earth that caused the KT extinction, including the extinction of the dinosaurs. The scientific team included: the father-son duo, physicist Luis W. Alvarez and his son, geologist Walter Alvarez; nuclear chemist Frank Asaro, and Helen Michel.

The primary global-scale kill mechanism was an impact-induced winter. According to this hypothesis, the impact of the gigantic asteroid blasted so much dust into the atmosphere that sunlight was blocked, and the Earth was plunged into an impact winter that triggered the KT extinction. The global blackout had a domino effect on plant and animal life existing at the time of the KT boundary. First, it killed off plant life that herbivores depended on for food. Then, as the herbivores died off, the carnivores that depended on the herbivores for food sources followed into extinction.

Many paleontologists were offended by the combination of the impact hypothesis for the KT extinction itself, and by the fact that it came from a group led by an outsider—a physicist for heaven's sake! As late as 1980, the prevailing view of the KT extinction among several prominent paleontologists was that the dinosaurs and other fossil groups that became extinct had been in decline long before the actual KT boundary, perhaps as long as 5 million years. They proposed that the floras and faunas of the late Cretaceous period gave way to new floras and faunas over a long time period—so long a time span that if an observer had been around for the whole time period, he or she most likely would not have understood what was happening. With tongue in cheek, the critics of the asteroid impact hypothesis would inquire how the dinosaurs and other species knew that an asteroid was about to strike, given that their decline had begun as much as millions of years before the impact.

The principal evidence that the Alvarez team had discovered to support their impact hypothesis was thin iridium-rich clay layers at locations in Italy and Denmark. The significance of the iridium find was that iridium is only present in the Earth's crust in extremely small amounts, but is present in much higher concentrations in asteroids, or meteorites. When a large impact occurs, the iridium in the extraterrestrial rock is vaporized, ejected into the atmosphere, and gradually falls back onto the Earth's surface through time. Rocks that were on the surface of the Earth at the time of the impact are then enriched in iridium, so when samples are examined in modern laboratories an "iridium spike" is observed. Additional support for the Alvarez team's 1980 hypothesis came from "iridium spikes" measured in globally distributed clay samples. An asteroid big enough to scatter the estimated amount of iridium observed in the worldwide "spike" was computed to have been about 6 miles in diameter and weighing in at 35 to 40 billion tons. Observations from recent craters and experimental results on impact dimensions from cannon shots indicated that craters are 10 to 15 times the diameter of the impacting object. This translated to an impact crater between 60 and 90 miles from the 6 mile asteroid computed by the Alvarez team.

Other evidence supporting the impact hypothesis for the KT extinction came from the discovery of shocked quartz and glass beads in many of the KT sections. Shocked quartz develops under high-pressure, high-temperature conditions—conditions that prevail during a high-velocity impact of an asteroid. The areal distribution of shocked quartz, high concentrations in North America low elsewhere, and glass beads in the southern United States and the Caribbean region, pointed to the latter region as the likely impact site. Then in 1991, in a relatively short paper (five pages) published in the *Geology* journal, a team of scientists led by a young Canadian graduate student announced the discovery of the location of the KT impact site. Based on data from boreholes drilled for oil exploration in the 1960s and later geophysical surveys, the team identified a crater centered on the village of Chicxulub on the Yucatan peninsula in southern Mexico. While the crater was buried beneath sediments deposited throughout the Tertiary period and could not be seen at the surface, the borehole and geophysical evidence suggested that it was approximately 90 miles across—remarkably close to the size predicted by the Alvarez team.

With the discovery of the Chicxulub crater in 1991, the hypothesis of an asteroid impact causing the KT extinction gained increasing acceptance among earth scientists to the point where many began to feel that the book was closed on this subject. As noted by the famous paleontologist Professor Michael J. Benton in his book *When Life Nearly Died*, "What a changed scientific world in the course of 20 years! In 1980, despite the work of the craterologists and the suggestion of a supernova explosion 65 mya, most earth scientists were still firmly in Charles Lyell's camp. When I learned my geology in the 1970s, my professors did not even mention impacts, craters, or mass extinctions."

By way of explanation, Charles Lyell (1797 to 1875) was a Scottish lawyer and geologist, and a very outspoken and influential advocate of the theory of Uniformitarianism, proposed a few decades earlier by the Scottish naturalist and founder of modern geology, James Hutton. This theory holds that geological processes at work in the present day are the same as those that operated in the past. Hutton, Lyell, and many geologists who followed in their footsteps between 1840 and 1980 espoused the theory of Uniformitarianism, and looked askew at anyone who even whispered the word "catastrophe." Such was the mood that confronted the Alvarez team and their 1980 hypothesis invoking a catastrophic cause for the KT boundary extinction rather than a multimillion year, more gradual, process of extinction of widespread species.

In addition to the less-than-enthusiastic reception that the Alvarez team faced from many scientists working on mass extinction issues, trouble was looming on other fronts for the impact hypothesis as the sole cause of the KT extinction. First, criticisms put forward by paleontologists in the 1980s were never addressed by the impact scenario. Many plant and animal species were found in the fossil record that were unaffected by the asteroid impact, and a successful extinction model has to account for this. In addition, many groups of organisms were in decline long before the impact, suggesting the workings of more gradual extinction mechanisms, prevailing at times well in advance of the KT boundary. Thus, all may not be lost for the Uniformitarians.

In March 1994, a group headed by a Princeton paleontologist, Gerta Keller, published a paper in the *Proceedings of the National Academy of Sciences,* titled "Chicxulub impact predates the KT boundary mass extinction." The authors presented evidence from a relatively deep borehole (more than 4,900 feet) drilled about 40 miles from the center of the crater. The borehole was drilled with the stated objectives to determine the precise age of the crater and its link to the global KT boundary layer and mass extinction. Keller found 20 inches of layered limestone in the crater that she said postdated the impact, but predated the extinction event. Keller suggested that the impact event preceded the mass extinction by about 300,000 years. She also reported finding foraminifers that had survived the initial impact, which, she believed, proved that the Chicxulub impact could not have caused the extinction of 75 percent of all the species, including the dinosaurs, at the end of the Cretaceous. While downplaying the Chicxulub impact as the KT culprit, Keller and her team hypothesized that an impact at the KT boundary, possibly associated with a crater in India and referred to as the Shiva crater, in combination with massive flows of basaltic lava ongoing for hundreds of thousands of years, was the straw that broke the camel's back, resulting in the demise of already fully stressed and diminished species. Needless to say, these results were highly controversial, and several prominent geologists and paleontologists went on record pointing out what they considered to be critical shortcomings of the Princeton group's research.

A huge outpouring of the Earth's interior that occurred 65 mya over much of present-day India came from deep in the Earth's mantle. How deep is a hotly debated issue with one camp of earth scientists favoring a plume of hot material originating at the core-mantle boundary, while another camp favors a more shallow starting point in the upper mantle? In either case, the outpouring of lava blanketed the India subcontinent with more than 250,000 cubic miles of basaltic lava. The stack of lava is known as the Deccan Traps. The word *traps* is derived from the Swedish word for stairs (trappa, or sometimes trap), and refers to the step-like hills or sheets of lava forming the regional landscape.

The Deccan Traps, one of the largest volcanic features on Earth, is referred to as a large igneous province, identified in Chapter 1 as a form of supervolcano distinct from the Toba explosive type. Given the location of the Traps on the Deccan Plateau of west-central India, it is properly referred to as a continental flood basalt. It consists of multiple layers of solidified flood basalt that together are more than 1 mile thick and cover an area of about 190,000 square miles. It's estimated that the original area covered by the lava flows was as large as 580,000 square miles, approximately half the size of modern India. The present volume of directly observable lava flows is about 123,000 cubic miles.

MOTHER OF MASS EXTINCTIONS

In 2005, two scientists (Vincent Courtillot and Thor Thordarson) pointed out that there is a strong correlation between the ages of massive flood basalts and the times of major mass extinctions. The correlation was first introduced around 1990, and has been reviewed several times since. The latest compilation of age data shows that each one of the last four major mass extinctions can be associated in time with a large flood basalt. This correlation suggests that a common mechanism may have played an important role in bringing about the mass extinctions. In order to examine this hypothesis, we will consider another flood basalt-mass extinction event. This flood basalt-mass extinction event is one that strains our vocabulary of superlatives. This is the Siberian Traps that coincided with the largest extinction in the history of the Earth: the end-Permian.

The Siberian Traps are the remnants of widespread volcanic activity that occurred in the northern portion of the supercontinent Pangea about 250 mya. Vast volumes of basaltic lava poured over a large expanse of primeval Siberia in a massive flood basalt event. Today, the area covered is about 770,000 square miles and estimates of the original coverage are as high as 2,700,000 square miles. The original total volume of lava is estimated to have ranged from 240,000 to 960,000 cubic miles, with individual eruptions of basalt lavas of 480 cubic miles or more.

Throughout the latter portion of the 20th century it was the opinion of some geologists that flood basalts could not have had a major impact on the global climate because their effusive style of eruption would not be powerful enough to eject gases into the stratosphere. In 2003, a team of Canadian and Russian scientists published a paper wherein they identified the Siberian flood-volcanic event as the most voluminous and explosive continental volcanic event known in the past 545 million years. They obtained ages for lavas corresponding to the lowermost and the near-uppermost units of 251.7 Ma and 251.1 Ma, respectively. Along with stratigraphic correlations and paleomagnetic evidence, these ages suggested that rapid extrusion of the entire approximate 4-mile thick composite sequence occurred in less than 1 million years. The time of extrusion was observed to coincide precisely with an age of 251.4 Ma, previously obtained for the end-Permian mass extinction, the most devastating biotic crisis known as the "Mother of Mass Extinctions." Finally, and most important for consideration of how the Siberian Traps might have contributed to the end-Permian extinction, extensive tuffs, pyroclastic deposits, and the presence of silicic volcanic rocks, such as rhyolite, all strongly suggested that a number of large explosive eruptions, possibly approaching supervolcano status, occurred during or before the eruptions of the basaltic lavas.

The Siberian Traps flood basalt is not the only candidate for the cause of the "Mother of Mass Extinctions" at the Permian-Triassic boundary. In February 2001, a team of scientists, lead by L. Becker, published a paper in *Science* announcing that they had discovered clear evidence that the end-Permian mass extinction, similar to the extinction at the KT boundary, was triggered by a collision with an asteroid. The scientists claimed to have found extraterrestrial helium and argon in rocks from the Permian-Triassic boundary in China and Japan. The gases were trapped inside fullerenes, or buckyballs, that are often associated with impact debris. When the team examined the helium and argon, they found isotope ratios unlike anything on Earth, but similar to those found in extraterrestrial rocks. On this basis, they argued that the fullerenes and their entrapped gases must have come from an impact. Later that same year, however, geochemists from other laboratories published results based on samples taken from the same boundary bed in China that apparently contained no buckyballs, helium, or argon. In fact, one Japanese geologist pointed out that the Japanese sample that Becker's team had analyzed didn't even include the Permian-Triassic boundary. All in all, while the evidence for an impact at the Permian-Triassic boundary is limited, and many geologists do not accept the impact hypothesis for the "Great Dying" at the end of the Permian period, it remains as one of the possible contributing causes of the extinction.

In 2005, scientists at the National Center for Atmospheric Research (NCAR) created a computer simulation showing Earth's climate in unprecedented detail at the time of the end-Permian extinction. The study gives support to the theory that an abrupt and dramatic increase in levels of carbon dioxide in the atmosphere triggered the massive die-off 251 mya. The rising temperatures in the atmosphere eventually affected ocean circulation, cutting off the supply of oxygen to lower depths and extinguishing most marine life. At the time of the end-Permian event, higher latitude temperatures were estimated to be 18 to 54°F higher than temperatures observed in modern times, and the massive flood basalts had released large amounts of carbon dioxide and sulfur dioxide into the atmosphere over a 700,000-year period. In order to study the effects of these extreme atmospheric conditions on the global climate and life in general, the National Center for Atmospheric Research

(NCAR) scientists used a state-of-the-art climate model that integrates changes in atmospheric temperatures with ocean temperatures and currents. The modeling results indicated that ocean waters warmed significantly at higher latitudes because of rising atmospheric levels of carbon dioxide, a greenhouse gas. The model predicted that the warming reached a depth of approximately 13,000 feet, shutting down the normal circulation process where cold surface water descends, taking oxygen and nutrients into the ocean depths. The implications of the study are that the elevated carbon dioxide is sufficient to lead to the demise of marine life, and ensuing high temperatures over land would eventually lead to the demise of large numbers of terrestrial life. The results of this study pointed out the importance of treating Earth's climate as a system involving physical, chemical, and biological processes in the atmosphere, oceans, and on land, all acting in an interactive manner.

In another study, Professor Paul Wignall, a British geologist, traveled to Greenland to investigate late-Permian stratigraphy to test the impact hypothesis as a cause of the end-Permian extinction event. The Permian extinction beds that Wignall examined lasted approximately 80,000 years and showed three distinctive phases in the plant and animal fossils they contained. Such a long process contradicted the catastrophic meteorite impact theory.

The most interesting finding in the Greenland rocks was a significant increase in carbon-12. The standard explanation (rotting vegetation) could not have caused such a marked effect. The answer to this puzzle came from a relatively unusual source: offshore drilling in the Gulf of Mexico, which tapped reserves of frozen methane hydrate from the ocean floor. Massive volumes of frozen methane hydrate, also known as methane clathrate or the "ice that burns," lie buried in the seabed around the world. Optimum conditions for the formation of clathrates exist where ocean water is more than 1,300 feet deep and bottom temperatures are below 34 to 35°F. Under these conditions, the methane hydrates are kept in a solid state by the combination of the hydrostatic pressure of the overlying column of water and the low temperatures of the deep water.

A geologist working with an offshore drilling outfit, who was familiar with the distribution of worldwide deposits of methane hydrate, recognized that sudden upwellings of the frozen deposits from the ocean floor (because of increased water temperatures) could be the source of the large quantities of carbon-12 observed in the Greenland rocks. Experiments to assess how large a rise in deep sea temperatures would be required to sublimate solid methane hydrate indicated that an increase of 9°F would be sufficient to cause the solid form of methane hydrate on the seafloor to gasify and ascend to the atmosphere.

When Wignall learned of the oil geologist's results, he used his carbon-12 data to estimate how much methane hydrate would have to be released to affect the isotope balance. Methane is one of the most potent greenhouse gases, and he deduced that unlocking frozen methane hydrate would have caused a temperature rise of 7 to 9°F over time. While not enough to kill off 95 percent of life on Earth, he realized this was a compounded effect. That is, a rise of about 9°F must already have occurred to prompt the frozen methane to melt. The combined temperature rise of 18°F is generally accepted as a figure capable of causing the massive loss of species from the world's oceans. Following the devastation in the oceans, the killing off of land-based species continued for thousands of years until, finally, fully 95 percent of the Earth's species were extinct. Thus, under this scenario there were two end-Permian killers: the massive Siberian flood basalts, and the release of methane hydrate from the sea floor.

In the October 2006 edition of *Scientific American*, Peter D. Ward, a paleontologist and professor in the biology department, specifically its Earth and space sciences division, at the University of Washington, described an extinction scenario for the end Permian and end Triassic events. Similar in some respects to the discussion in the preceding paragraphs, his scenario does not involve asteroid or meteorite impacts, but does involve massive flood basalts, global warming, and an oxygen depleted ocean, at which point the similarity of scenarios ends.

As Professor Ward concludes in the October 2006 article:

But the most critical factor seems to have been the oceans. Heating makes it harder for water to absorb oxygen from the atmosphere; thus, if ancient volcanism raised CO_2 and lowered the amount of oxygen in the atmosphere, and global warming made it more difficult for the remaining oxygen to penetrate the oceans, conditions would have become amenable for the deep-sea anaerobic bacteria to generate massive upwellings of H_2S. Oxygen-breathing ocean life would have been hit first and hardest, whereas the photosynthetic green and purple H_2S-consuming bacteria would have been able to thrive at the surface of the anoxic ocean. As the H_2S gas choked creatures on land and eroded the planet's protective shield *(reference here is to the ozone layer)*, virtually no form of life on the earth was safe.

A poisoned ocean and atmosphere would also explain the slow recovery of life after the mass extinction.

On October 25, 2006, two scientists introduced a hypothesis at the Annual Meeting of the Geological Society of America that considered what the different extinction mechanisms that have been proposed over the years have in common. This new hypothesis, referred to as the Press-Pulse hypothesis, rejects the all-or-nothing approach to mass extinction, and calls on a combination of sudden and deadly catastrophes. The "pulses" correspond to impact events. The "presses" correspond to the longer, steadier pressures on species, which include massive flood basalts and ensuing global effects.

To test the Press-Pulse hypothesis, the scientists compiled extinction rates for the past 488 million years. As previously noted, impact cratering events served as a proxy for pulse disturbances, and continental flood basalt events stood in for press disturbances. What they found was that average extinction rates were similar during geologic periods of time when either press or pulse events occurred alone. Extinction rates during these times were statistically indistinguishable from rates associated with time periods when neither impacts nor flood basalts occurred. In contrast, when press and pulse events occurred together, higher average extinction rates were observed. They also noted that the size of the associated flood basalt or cratering event was poorly correlated with extinction rate. Thus, they concluded that it is the combination of press and pulse events—a geologic one-two punch—rather than the magnitude of single events that explains Earth's greatest episodes of extinction.

In previous sections of this chapter and, Chapters 3, 5, and 9, there were discussions of the impacts on the environment and humans of a depleted stratospheric ozone layer. An increase in the UV-B radiation, which is normally shielded by the stratospheric ozone layer, would ultimately have a very deadly effect on land plants and animals. In 2004, a group of scientists from the Netherlands, the United States, and the United Kingdom published the results of a study of land plant mutations that occurred at the end of the Permian period. These authors observed that during the end-Permian ecological crisis, terrestrial ecosystems experienced preferential dieback of woody vegetation. They attributed the dieback to exposure of the plants to enhanced UV-B radiation which, in turn, was likely a consequence of severe disruption of the stratospheric ozone balance by massive emissions of hydrothermal organohalogens in the vast area of the Siberian Traps. While extraterrestrial scenarios, such as supernova and gamma ray bursts, have been suggested in the past to account for depletion of the stratospheric ozone at times of mass extinctions, the preponderance of evidence favors a more mundane scenario, augmented perhaps by asteroids or meteorites from space.

We conclude this chapter with a discussion of an all-in-one mass extinction mechanism that is best described as a supervolcano extraordinaire. In 2004, a team of three scientists at Kiel University in Germany published an article in *Earth and Planetary Science Letters* that defined the mechanism as a "Verneshot" after the French author, Jules Verne, who pioneered the science-fiction genre. The hypothetical mechanism appears to have one big advantage over many of the others proposed to date. Namely, it explains a seeming mystery that plagues the continuing debate over the four major mass extinctions: why the extinctions always seem to coincide with both continental flood basalts and asteroid, or meteorite, impacts when the odds of these two events happening simultaneously four times are very small. The authors calculate a probability of four coincidences happening within the past 400 million years of 1 in 3,500.

The first idea the authors considered was whether impacts could cause continental flood basalts. They very quickly determined that it was likely physically impossible for an impact to cause the sustained melting associated with a continental flood basalt. At this point they decided to take an entirely different approach to the problem. If impacts could not trigger the prolonged outpouring of the continental flood basalts, could flood basalts somehow generate the geological and geophysical "signals" characteristic of extraterrestrial impacts (that is, craters, iridium anomaly, shocked quartz, and buckyballs). While this idea was mentioned as early as the 1960s, no one had, as of 2004, come up with a plausible mechanism to explain how flood basalts could produce the "signals" of an impact.

The hypothesis that the three scientists came up with involves a mantle plume welling up under a craton, which is an old and stable part of the continental crust that has a thick crust and a root that can extend 125 miles into the mantle. Because of the crustal thickness, the plume would not be able to break through at first, but would stall at a depth around 50 miles, melt carbon-rich rocks, and result in an out-gassing of large quantities of carbon dioxide and sulfur dioxide. The final straw that breaks the camel's back, or causes the 50 mile thick cap to blow, is a rifting of the craton. The authors state that there is evidence that rifting was going on during the three most recent extinctions: the end-Permian, Triassic, and Cretaceous. The rifting would be enough to release the pressure inside the craton and enable a catastrophic gas explosion. But immediately after the explosion, the pressure would plummet in the pipe that carried the gases to the surface, causing it to cave in from the bottom upwards. The sudden carbon dioxide/sulfur dioxide release into the atmosphere would provide the primary killing mechanism of the induced extinction event. Such explosive, deep, lithospheric blasts could blast out rocks from the top of the pipe into suborbital trajectories to subsequently impact the Earth's surface at distant locations.

The tantalizing aspects of this hypothesis are that it accounts for all the impact "signals" mentioned previously. In an interview with the lead author of the 2004 paper, Phipps Morgan related how the proposed mechanism reminded him of a book he read as a child. The book was Jules Verne's *From the Earth to the Moon*, which is about a huge gun that shoots objects into space. Phipps commented, "we decided to name our mechanism after Jules Verne's space gun." Thus, the verneshot hypothesis was born.

How has this hypothesis been received by the geological community? The responses have varied from, "It's a creative approach to a real problem," according to Harvard scientist Paul Hoffman; to "there is not a scrap of evidence for Verneshot events," a quote from scientist Jan Smit at the Free University of Amsterdam. A sharp, but interesting, criticism came from the Belgium scientist, Philippe Claeys, who pointed out that, in his opinion, only the KT extinction event had clear evidence of an impact. He suggested that the verneshot originators consider the simpler option that the impact signals at boundaries other than the KT are spurious. "If that is the case we don't need any mystic and un-testable mega-volcanic hypothesis to solve the problem."

There is one more mass extinction event that we have not yet considered. This is an event that could even surpass the Big Five mass extinctions of the past 450 million years. This event is referred to as the "Sixth Extinction," and it's happening now. Scientists estimate that during the 20th century, between 20,000 and 2 million species became extinct. While the total number cannot be determined more closely, studies conducted during the past 50 years are yielding a better estimate of the current rate of extinction. In his book, published in 1995 with coauthor Roger Lewin, Dr. Richard Leakey, one of the world's most famous paleoanthropologists, stated that the current extinction rate is between 17,000 and 100,000 species per year. Assuming the number is 50,000 per year, results are that 50 percent of the Earth's species will be killed off before the end of this century.

What is the kill mechanism? There are no massive continental flood basalts erupting. No asteroids of significant size have impacted the Earth in recorded history. There have been no verneshots, no massive gamma ray bursts or supernovas. Pogo, the cartoon character created by Walt Kelly, nailed it when he responded to Porkypine's question about the garbage strewn Okefenokee Swamp in Georgia, "Yep, Son, we have met the enemy, and he is us."

CHAPTER 8

The Psychology of Survival: Mass Trauma and Cellular Memory

Toba's eruption left an undeniable imprint on the genetic makeup of every human being that has existed since. The physical evidence, seen in the tiny mitochondrial DNA, tells the story of a population bottleneck, which almost wiped out humanity. But that story is not complete without also looking at the psychological imprint that a catastrophic event might possibly leave upon any survivors. Could cells, and the DNA that exists within each cell, harbor the imprint, passed down from generation to generation, of a trauma so severe and widespread, that it literally changed the physiology of the survivors and their subsequent generations of descendants?

Blue eyes and blonde hair are handed down via DNA, but could psychological traits such as depression, anxiety, and phobias also be delivered to future generations via the intricate chain of cellular structure?

Millions of people each year are affected by catastrophic events such as floods, tsunamis, hurricanes, earthquakes, and volcanic eruptions. Natural disasters spell instant death for many, but for survivors, long-term post-traumatic stress disorder and negative psychological disorders affect them for the rest of their lives, even if they are lucky enough to afford a good therapist.

Imagine, though, a trauma so severe that it literally brings humanity to the brink of extinction. Those who live to witness the mass death and destruction, and face the ongoing struggle for survival in a world totally different from the one they've known, are prime targets for the kind of mass psychological damage that some scientists are beginning to think may be imprinted upon the very DNA that makes each one of us who we are.

Normally, a natural disaster brings out both the best and worst in people. Similar to the terrorist attacks of September 11, 2001 in America, disasters shock and numb the populace, yet also serve as catalysts for bringing people together and creating an environment for positive change.

If we look at events such as September 11th, the Indonesian earthquake and tsunami, the Columbine school shootings, and Hurricane Katrina, we see patterns in stress reactions that occur in both children and adults over the following days and weeks. According to the National Center for Post Traumatic Stress Disorder, those reactions include:

✳ **Emotional Reactions:** Shock, fear, grief, anger, guilt (especially survivor guilt), shame, rage, helplessness, hopelessness, emotional numbness.

✳ **Cognitive Reactions:** Confusion, disorientation, indecisiveness, worry, shorter attention span, difficulty concentrating, memory loss, unwanted memory, self-blame.

✳ **Physical Reactions:** Tension, fatigue, difficulty sleeping, body aches and pains, edginess, easily startled, nausea, change in appetite and sex drive, rapid heartbeat.

✳ **Interpersonal Reactions:** Distrust, irritability, conflict, withdrawal, isolation, feeling judgmental or controlling, feeling rejected or abandoned.

These reactions are common and often dissipate over time, but the National Center for PTSD also states that one in three survivors will not cope as well, and will suffer from lasting PTSD, which includes symptoms of:

* **Dissociation:** Nothing feels real, feeling outside of yourself, blank periods with no memory.

* **Intrusive Reexperiencing:** Terrifying memories, flashbacks, and nightmares.

* **Avoidance:** Extreme attempts to avoid memories, through substance abuse or addictive behaviors.

* **Emotional Numbing:** Unable to feel emotion, has a feeling of emptiness.

* **Hyperarousal:** Panic attacks, rage, intense irritability, and agitation.

* **Severe Anxiety:** Paralyzing worry, extreme helplessness, compulsions, or obsessions.

* **Severe Depression:** Loss of hope and the will to live.

The greater the trauma, the more severe the readjustment problems the survivors face. If the traumatic event includes factors such as witnessing widespread death; life-threatening danger or harm (especially to children); loss of home and community; extended exposure to danger, loss, and environmental stressors; severe hunger and lack of shelter and water; and exposure to toxic contamination, then PTSD becomes a lasting problem with deep impact upon the body, mind, and spirit. Severe trauma often leads to addictive behavior as a means of coping and self-medication, which can often prolong the original trauma. Without effective means for dealing with the trauma, survivors run the risk of suicide and extreme behavioral changes.

The Federal Emergency Management Agency's (FEMA) literature on surviving natural disasters states that everyone is deeply affected by a catastrophe on one level or another, and children and the elderly or disabled have even bigger challenges. But in today's modern society, even with mistakes made during Hurricane Katrina still making headlines, most of us can rest assured that help will be on the way, or that we can get to the nearest shelter.

TOBR'S EMOTIONAL DEVASTATION

But imagine a much more dire situation approximately 74,000 years ago.

The eruption of Toba certainly included drastic and ongoing stressors such as those previously mentioned, amplified to an extreme, and no doubt changed the way survivors functioned on the most basic, fundamental levels. These survivors did not have the Red Cross and FEMA, cell phones and ham radios, or mass transportation, emergency shelters, and food drops via helicopters. They did not have an Emergency Broadcast System to warn them to evacuate, or a police force or fire battalion to help guide them to safety. They had nothing but themselves and the land, and the land itself was ravaged from the ensuing volcanic winter. All around them, members of their family, tribe, and makeshift community were falling victim to famine, contaminated water supplies, disease, and possibly, even the progressive effects of fatal doses of ultraviolet radiation pouring through a depleted stratospheric ozone layer.

Almost extinct.

Situational trauma, or that trauma brought about by an event or disaster such as a supereruption, may be so overwhelming to the survivors that coping mechanisms break down altogether, especially if no qualified help is available for dealing with the severity of the situation. Some people react to traumatic events, while others become proactive and take charge, and still others go numb and do nothing in extreme situations. Trauma can also be divided into two distinctions involving time. Recent trauma is considered "simple" trauma, and trauma from past or distant events is known as "complex" trauma, each with its own different coping styles and methods.

For the survivors of Toba, the trauma of the eruption was immediate, yet also long lasting as the volcanic winter set in, and starvation and lack of shelter became the immediate challenges. The survivors experienced continual stressors as they struggled to stay alive and keep the species viable. They may have been experiencing what neurologist Jean-Martin Charcot called "traumatic hysteria," a condition rendering them physically and emotionally paralyzed, unable to deal with the rapid changes around them, even years after the "incubation" period following the initial event. Charcot believes that psychological trauma is the origin of all hysteria.

In times of war, we see cases of PTSD, usually following bouts of acute stress disorder, or ASD, in many returning veterans. Often, the ASD and PTSD does not manifest for years after they have returned from the front lines. The psychological damage of war is comparable to what the survivors of Toba may have felt, watching their loved ones die, seeing the land around them decimated, and yet, for those who lived through Toba, there may have been no home to go back to.

Collective trauma, or mass trauma, can be the most damaging. Events such as the September 11, 2001 terror attacks, the John F. Kennedy assassination, and the 2004 Indonesian earthquake/tsunami can literally change the mental and emotional mindset of entire communities, and even nations. Collective traumas are often watershed moments in historical and societal change, sometimes resulting in revolutions, mass protests, civil wars, and uprisings against oppressive governments.

Often, the actual identity of a group of people is altered, as in the Law of Common Fate, where people who share a traumatizing experience associate themselves with a certain identifier, and all others are treated as outcasts. This occurred during World War II when the submarine *USS Puffer* was under attack by a Japanese surface vessel. During the hours of the terrifying attack, the crewmen bonded in a psychological way that only became obvious later, when other crewmen who were transferred to the *Puffer* were treated as outcasts.

CELLULAR MEMORY

Shared experience is contagious to a certain degree, bonding even those who suffer tremendous loss during the experience. The collective consciousness of survivors shifts, creating groupthink, or herd behavior, drastically altering the behavioral patterns of large groups of people. But does collective trauma affect humans on a fundamentally physical, even potentially genetic level? Does our shared grief, fear, and anguish change us at the cellular level?

Tian Dayton, Ph.D., writes in *Trauma and Addiction: Ending the Cycle of Pain Through Emotional Literacy* that "trauma, by its very nature, renders a person emotionally illiterate." This emotional breakdown cripples the person's ability to sense reality, and respond to it accordingly. Dayton's book points to scientific research in the field of neurobiology that shows through brain imaging that trauma can affect the body and the brain in ways we've never understood before. In fact, traumatic memories may be stored throughout the body as what many scientists are now calling *cellular memory*.

Dayton suggests that one of the first scientists to come up with cellular memory was Jacob Levy Moreno (1889 to 1974), the creator of the psychodrama. Moreno believed that "the body remembers what the mind forgets," and he and his wife, Zerka, who co-developed psychodrama during the latter half of the 1920s, used this form of psychotherapy as a means to reach those lost memories. Moreno, who published *The First Book on Group Psychotherapy* in 1931, felt that there were two types of actual memory:

Content Memory: Thoughts, recollections, feelings, facts.

Action Memory: Tensions, holding, warmth, tingling, incipient movement.

Neuroscientist Candace Pert suggests that "intelligence is located not only in the brain, but in cells that are distributed throughout the body....The memory of trauma is stored by changes at the level of the neuropeptide receptor...." In her book, *Molecules of Emotion* (Simon and Schuster), Pert states that peptides and other informational substances are "biochemicals of emotion," and that emotional expression is tied to specific flows of peptides in the body, with those involving serious trauma resulting in severe disturbances to the psychosomatic network of the body. Pert also believes this traumatic disturbance can be stored in a specific body part.

In other words, our cells may have memory of past events, including severe trauma, just as they hold the memory of our genetic makeup.

In a paper for the American Academy of Experts in Traumatic Stress, titled "Decoding Traumatic Memory Patterns at the Cellular Level," Thomas R. McClaskey writes that, "Virtually every behavioral pattern exhibited during routine activity of daily living results from learned data which is stored, or encoded, as cellular memory." While most of these patterns are not profound enough to contribute to disease or illness, McClaskey suggests that certain patterns are "expressed as significant reflections of traumatically encoded cellular information."

Pointing to the famous work of Russian psychologist Ian Petrovich Pavlov, who won a Nobel Prize for his work on the digestive process, McClaskey reminds us that Pavlov's conditioned reflex responses are associated with memory. During the shock or stress of the event, which is perceived as a threat either physically or emotionally, a complex of hormonal messenger molecules are released by the limbic-hypothalamic-pituitary-adrenal system. These hormonal soups, so to speak, encode the external and internal threat of the event as cellular memory. The initial stimulus, McClaskey continues, serves as the catalyst to later produce the conditioned response and basic reaction, such as Pavlov's dogs salivating at the sound of a bell, thinking they will be fed.

McClaskey concludes that all memory is encoded at the cellular level. But his suggestions for various forms of therapeutic treatment simply would not apply to those survivors of a primitive culture, and their ensuing generations. The psychosomatic and psychological damage of Toba upon the survivors would have been more than they could possibly deal with or understand. And we, the ultimate descendants, might still be feeling the impact of that collective trauma today, on a foundational, cellular level.

One has to only close one's eyes and imagine what the Toba survivors must have experienced, from the terror of the initial supereruption to the after-effects: months on end of daylight reduced by approximately 75 percent and nights without starlight, the moon dimmed by the atmospheric ash and gas-ses, temperatures plunging to freezing at night, lack of food, water, and shel-ter. UV-B radiation would take its toll on their own bodies, even as they watched their livestock perish. The survivors would have no scientific knowl-edge to explain what was happening, and would exist solely from a position of total ignorance and fear. Fear of the wrath of the Gods, fear of nature, fear for their own physical, mental, and even spiritual sanity, fear for their very survival.

Fear can alter birth weight. Several recent studies, published in the 2006 *Journal of Psychosomatic Research*, show that pregnant women who gave birth shortly after the September 11th terrorist attacks suffered greater stress that resulted in both an increased rate of stillborn males, and lower birth weights overall. Dutch scientists at Maastricht University conducted a study with 1,885 women who were at least 12 weeks pregnant at the time of the attacks. Their findings suggest increased levels of the stress hormone cortisol as one of the culprits. Cortisol can transfer from mother to fetus, and higher levels result in lower birth weight by reducing blood flow to the fetus and potentially stunting their growth.

The body does respond to fear, and that fear is passed to the next generation in the form of physiological effects. But can a traumatic event become embedded in the human body? Do cells remember? That is the question many psychologists and geneticists are trying to answer, and some of the most compelling evidence has come from a very unusual source—organ donors. People who have received a heart, lungs, and other major organs from donors often report unusual behaviors, cravings, and even memories that they claim were not theirs to begin with. Similar to the genetic data codes that make us look the way we do, the body's cells may contain another type of code that makes us act the way we do.

In a May-1998 special report titled "Do Cells Remember?" for *USA Weekend.com*, reporter Pam Janis presented the case of Claire Sylvia, a heart-lung transplant recipient who claimed to have developed unusual cravings for beer. She had never liked beer...until she received the organs of an 18-year-old named Tim, who loved it. Sylvia's story has been documented in a book and movie called *Change of Heart* (Warner Books). Janis also points to the work of psychologist Paul Pearsall, a bone marrow transplant recipient who's own book *The Heart's Code* (Broadway Books) discusses his belief that the heart is loaded with information the brain should not dismiss.

"The brain has lost its mind," Pearsall states. "The heart has a coded, subtle knowledge connecting us to everything and everyone around us." But not all scientists agree that transplanted organs can carry with them coded information. Some, similar to John Schroeder, a cardiologist at Stanford Medical Center, believe the concept is unimaginable and insists that "psychological experience is stored in the brain." Jeff Punch, M.D., a transplant surgeon at the University of Michigan, believes that the entire concept of cellular memory is "supernatural," and that organs are "not capable of transferring memory to a person's mind in any conventional sense," as he stated in an article titled "Cellular Memory" on *TransWeb.org*.

But Pearsall argues that science needs to be more open to the idea that the heart is a sentient, thinking, and communicating organ. The "theory of neuro-transmitted emotion" has been researched heavily by neuroscientists such as Candace Pert. In *Molecules of Emotion*, she reminds us that neuropeptides, those biochemicals of emotion we talked about earlier, are found all over the human body, and that memories are held in the receptors of these neuropeptide "ligands." When a specific receptor is flooded with a specific ligand (via the release of emotion/s stored in the body), the cell membrane is altered, either inhibiting or facilitating the choice of neuronal circuitry that will be used. This change in neuronal circuitry then changes the emotion initiated, according to Pert. Pert also points to a study by Miles Herkanham that shows less than 2 percent of neuronal communication occurs at the synapse, suggesting that all atomic, molecular, and cellular systems may play a role in the storing of information and energy.

Interestingly, this same neuronal communication on a cellular level could explain why some organ donors reject their organs. Perhaps the body not only rejects the actual material organ, but also the stored memory and information held at the cellular level. Extreme or chronically suppressed emotions, such as those resulting from a catastrophic event, may be too overwhelming to the organ, as well as the actual transplanted donor cells.

In a study Pearsall conducted with University of Arizona scientists and authors of "The Living Energy Universe," Gary Schwartz, Ph.D., and Linda Russek, Ph.D., published in the Spring 2002 issue of the *Journal of Near-Death Studies*, the three researchers conducted open-ended interviews with 10 heart- or heart-lung-transplant recipients, their families and friends, and the friends and families of the donors. The researchers found some very striking examples of the recipients experiencing changes in behavior, personality traits, and likes/dislikes.

Some examples from that study are:

* A 7-month-old boy received a heart from a 16-month-old drowning victim who had a mild form of cerebral palsy on the left side of his body. The recipient, who had no indications of CP, went on to develop similar symptoms, such as shaking and spasticity on the left side.

* A 29-year-old lesbian and fast-food lover received the heart of a female 19-year-old-boy-crazy vegetarian. The recipient found herself unable to stomach meat after the operation, and became engaged to a man, no longer finding women attractive.

* An 8-year-old girl received the heart of a 10-year-old murder victim. Afterward, the recipient experienced horrible nightmares

of a man murdering the donor girl. These dreams so traumatized the recipient that psychological help was sought, and police later used actual details the recipient gave when asked to describe her nightmares to help catch the donor's murderer.

The three researchers propose a living systems theory, which posits that living cells possess memory and decider functional subsystems within, and that the recent integration of systems theory with the concept of energy, known as "dynamical energy systems," allows for the prediction that all dynamical systems store information and energy in varying degrees, not just the brain system or the heart system. Thus, atomic, molecular, and cellular systemic memory should be possible.

The 10 cases examined in the study were representative of more than 74 transplant patients that Pearsall examined throughout a 10-year period. The three researchers are conducting further studies at the University of Arizona involving 300 transplant patients, using semi-structured interviews and systematic questions, which they hope will cast further light on a subject of growing interest, and controversy. As more recipients and donors are willing to be interviewed, and openly discuss personality issues and changes, more potential exists for one day proving, or disproving, the theory of cellular memory transfer.

Most scientists will explain these cases, and dozens of others similar to them, as just coincidences, or "bizarre psychological aberrations." Other scientists blame the toxic effects of transplant drugs and associated surgical trauma, or the sheer psychological implications of receiving another person's vital organ that had to die for their own life be saved. But some, such as Dr. Jack G. Copeland, University Medical Center chief of cardiothoracic surgery in Arizona, suggest there is something to the phenomenon, although it may happen to a small minority of patients. In an article for the *Arizona Daily Star* called "Transplant Memory?," Copeland, who is part of a team that has performed more than 700 transplants during the past 25 years, told reporter Carla McClain, "It's highly controversial, but I don't exclude it completely." The article quotes Dr. Sharon Hunt, heart transplant surgeon at Stanford University Medical Center, as calling cellular memory "Fiction….There is no science to explain such a thing."

But Copeland does not completely dismiss the notion, citing that "with any solid organ, you are transferring DNA from the donor to the recipient…these are genes that relate not only to the specific organ, but to other systems as well, such as cerebral function. So there may be something to this thing that personalities can change."

Still others argue that, if every cell contains within it the entire DNA blueprint, why then could every cell not also contain the stored information of memory? According to Joseph Chilton Pearce, in an intriguing interview with Chris Mercogliano and Kim Debus for the 1999 *Journal of Family Life*, "the idea that we can think with our hearts is no longer just a metaphor, but is, in fact, a very real phenomenon." Pearce, a noted scholar, author, and scientist, discusses the amazing ability of the heart to do what the brain has always been given credit for: think.

THE HEART

In fact, he suggests the heart is the major center of intelligence in the human body, and that, quite literally, there is a "brain in the heart." The heart is the major endocrinal gland of the human body, producing a hormone called ANF (Atriol Neuriatic Factor) that affects the limbic brain, or the emotional brain, including the hippocampal region where memory is said to reside. According to neurocardiologists, approximately 65 percent of the heart's cells are neural cells, just like those of the brain, operating through the same connecting ganglia, axonal, and dendritic connections as in the brain.

The responses made by the heart effect the entire human system, and the heart is also a very powerful electromagnetic generator, creating a field that encompasses the body, extending out 8 to 12 feet away. Pearce states that the field the heart produces is holographic, and that you can read this field from any point on the body, or from any point within the field itself. No matter how microscopic a "piece" of the field you pick, you can read the whole information of the entire field.

The electromagnetic spectrum of the heart is also profoundly affected, according to Pearce, by our emotional responses to our environment, and this is what the brain uses for information. "Ultimately, everything in our lives hinges on our emotional response to specific events." Pearce also points to research in England involving profound environmental changes and their effects on the genetic structuring within the body. These studies suggest our DNA is not locked into "unchanging programs, as previously thought, but in fact are profoundly affected by our environment, particularly our emotional environment." He points further to research done with pregnant women, and how the emotional state of the mother during pregnancy determines the evolutionary direction within the developing fetus. The mother's well-being determines if the development will eventually be concentrated in the frontal lobes, where intellect and creativity are centered, or in the ancient, survival-based reptilian brain. Thus, our emotions have a direct role in our evolutionary development.

"This is probably the most explosive information to come along in quite a while," Pearce states. "And this makes perfect sense because the heart is the first organ to form in the fetus…and it has to be because it furnishes the electromagnetic spectrum upon which DNA itself depends for its instructions."

The concept of cellular memory is not new. Atillio D'Alberto, Doctor of Chinese Medicine in Beijing, China, and author of *Cellular Memory and ZangFu Theory*, examined the oldest and most important medical book from China, the *Huang Di Nei Jing* (*Yellow Emperor's Internal Medicine Classic*), and found many correlations in one specific part of a Five-Element Theory, called ZangFu. The ZangFu (Organ and Viscus) element involves pathology and physiology, and consists of five Yin (Zang, or solid) organs, and five Yang (fu, or hollow) organs. Each organ has a function, or element, and each organ has its own corresponding emotion and spirit.

D'Alberto discusses the role of yin and yang theory: that everything is made of two opposing forces, each containing the seed of its opposite. This holistic, holographic view suggests that there is cellular communication with the "body-mind in dynamic interplay," and that every piece of the whole can be seen in the cellular structure of all living bodies. Therefore, each organ contains the functional essence of all the characteristics of the ZangFu organs, and the entire body as a whole.

To take it one step further, each organ "houses" the seed or essence of all the other organs' emotions and spirit within the body. D'Alberto compares it to a computer: the ZangFu are the hardware; emotions and belief systems are the software; the mind or *Shen* is the operating system; and the brain is the microchip.

More recently, Jungian psychology introduced the concept of "racial memory" or "genetic memory," which hypothesizes that an organism's accumulated knowledge and experience can be carried genetically into its descendants. The idea is that this carried information is somehow hardwired into the offspring, even though no actual person-to-person contact or communication has occurred.

Carl Jung's psychological paradigm suggested that racial memory is inherited genetically, as part of a collective unconscious of the human species that does not require immediate experience or conditioning. But many modern scientists and psychologists have discredited the idea of racial memory, considering it more fiction than truth. However, some laboratory experiments involving female mice have shown that a female under extreme stress will produce offspring with the biochemical makeup for higher levels of stress and anxiety. There are also the stories of the common chimpanzees who demonstrate a distinct skill their ancestors mastered, even when isolated from direct contact.

Ancestral memory, or genetic memory, is potentially like "factory-installed software," according to Darold Treffert, M.D., Clinical Professor for the University of Wisconsin Medical School's Department of Psychiatry, who stated in an article for the Wisconsin Medical Society Website that savants, those prodigious geniuses who possess amazing musical or intellectual skills in the absence of formal training or exposure, could possibly be accessing "the genetic transmission of sophisticated knowledge (beyond instincts)." Treffert again points to the collective unconscious of Carl Jung, but also to the work of Wilder Penfield, who wrote in his pioneering book, *Mystery of the Mind*, that there were three types of memory. "Animals," Penfield states, "particularly show evidence of what might be called racial memory." The other two types of memory he cites are conditioned reflexes and experiential. Only racial memory does not require contact or experience in the present.

Perhaps this racial, or genetic memory, is the result of "factory-installed software," as Michael Gazzaniga suggests in his 1998 book, *The Mind's Past*, when he says "As soon as the brain is built, it starts to express what it knows, what it comes with from the factory. And the brain comes loaded." He goes on to say that, "these things are not learned; they are innately structured."

This could explain how savants and prodigies know what they know, despite the fact that they couldn't possibly, on the surface, know what they know. It's as if the savant comes "already programmed with a vast amount of innate skill and knowledge in his or her area of expertise," states the National Heart Institute's Marshall Nivenberg, in an article titled "Genetic Memory" in the 1968 edition of the *Journal of the American Medical Association*

(JAMA). Nivenberg also points to the "massive cognitive and other learning handicaps" savants often suffer, and yet still they know what they know and are able to excel at certain tasks and skills.

EVOLUTIONARY PSYCHOLOGY

Evolutionary psychology (EP) might also lend some credence to the possibility that our ancestors passed their memories down to us genetically. EP is based upon the theory that there is a genetic basis to the functional structure of cognition, just as there is for the structure of specific organs and bodily systems; it also focuses on the evolved properties of the nervous system. This functional structure serves the purpose of solving survival and reproduction problems in a particular species, divided into two types of adaptive problems:

✱ Recurrent adaptive problems: Those that occur consistently over the course of a species evolutionary history.

✱ Novel adaptive problems: Those that occur spontaneously or erratically.

Many experts agree that little is known about the Pleistocene, the era in which humans developed in an evolutionary sense. Because of this, there is controversy as to whether evolutionary psychology can accurately depict the environment and environmental features present during the time that we, as a species, most evolved. But many evolutionary psychologists claim that we know enough about the challenges of our hunter-gatherer ancestors to understand the necessary adaptations required for survival, including predator and prey, child rearing, mate choice, food acquisition, diseases, and, of course, environmental factors such as volcanic eruptions that created significant natural selection pressures.

Interestingly, one example evolutionary psychologists use to show that human psychology is indeed adapted to the Pleistocene environment of evolutionary adaptedness, or EAA, comes from our fear of spiders and snakes. Cars kill more than 40,000 people every year in the United States alone, yet spiders and snakes kill maybe a handful. But people more readily learn a fear of spiders and snakes than they do cars, guns, or other dangers that kill far more. One explanation of this is that spiders and snakes were a major threat to our ancestors, living as close to the earth as they did. The fear of snakes is indicative of the predictions evolutionary psychologists make in terms of "attachment theory," which claims that an organism or mechanism will adapt to interact with its surroundings. Apparently, we humans are more adapted to the Pleistocene and all its apparent dangers than to our own modern day environment.

Because EP is so focused entirely on issues of survival and reproduction, a bottleneck that reduced the human population so drastically, as Toba's eruption did, must have no doubt created major stressors to the natural selection process of humans and other animals. We know that major changes to an environment can cause genetic mutations, such as exposure to toxic chemicals or gasses, radiation, and so on. The vast majority of DNA mutations will result in changes to the body that directly affect reproduction, the sole means of a species' survival. Evolutionary time, or the span of time necessary for mutations to arise and spread in the population, is approximately 1000 to 10,000 generations. This equates to 20,000 to 200,000 years, a significant amount of time for a species to develop or adapt to their environmental stressors. It's interesting that, in order to understand the brain and how it works in a cognitive sense, we have to look back to the environment of our ancestors and examine what was happening to them.

Some may ask why the Pleistocene era is so crucial to our human development. This span of time begins 1.8 million years ago and ends approximately 11,000 years ago. It is considered a comfortable amount of time, according to Edward H. Hagen of the Institute for Theoretical Biology in Berlin, to allow complex adaptations to evolve. The main time focus falls between 200,000 and 20,000 years ago, well within the Pleistocene, and includes the time toward the end of the era when humans moved from hunting and gathering to agriculture.

The Pleistocene encompasses the vast majority of the origins of our genus, but, as Hagen points out, excludes the most recent 10,000 years, during which there has been tremendous development, allowing for novel selection pressures to arise. "The reason we can still roughly equate the Pleistocene with the period of time which shaped human adaptations is that, if an adaptation which evolved prior to the Pleistocene, like vision, were not under stabilizing selection during the Pleistocene, that adaptation would have been lost during the 1.8 million years of the Pleistocene," Hagen explains in *The Evolutionary Psychology FAQ*.

The physical adaptations of natural selection, due to the environmental factors present, are obvious to us. But again, what about the psychological adaptations of a small group of humans forced to survive an event so brutal and horrific? Some scientists question if there are even enough genes present in humans to account for psychological adaptations, pointing to the fact that the human genome contains only 30,000 or so genes, although some current estimates put that number between 20,000 and 60,000. Hagen suggests that we think of adaptations not as simple product of genes, but as products of gene interactions. "It is well-known that both gene and non-gene regions of DNA control the protein production of other genes, and that multiple proteins combine to produce an adaptation," Hagen states. Therefore, for an organism with, say, only 30,000 genes, we can then calculate that the number of gene interactions, or two-gene combinations, goes up to almost half a billion. The total number of a 25-gene combination ends up being greater than the number of particles in the universe (possibly excluding dark matter)!!!

"An organism obviously need make use of only a minute fraction of such gene combinations to produce an incredibly rich, functionally organized phenotype with enormous number of adaptations," Hagen concludes.

The idea that psychological adaptation can occur just the same as physiological adaptations, and that cellular memory may be passed down from our ancestors in the form of those adaptations, can find some credence in recent studies about the role genetics plays in depression, fear, and anxiety. It has long been concluded that depression, and some forms of anxiety disorders, run in families. Genes can determine our hair and eye color, and can also determine some of the illnesses we might inherit from familial lines. Research into the heredity of depression and anxiety suggests that some individuals are more prone to develop these illnesses than others.

In an article titled "Genetic Causes of Depression" for *AllAboutDepression.com*, bipolar disorder is said to have a strong genetic influence, with approximately 50 percent of those diagnosed having a parent with a history of clinical depression. If both parents have clinical depression, the percentage rate of a child developing bipolar disorder is between 50 and

75 percent. Even having a sibling with bipolar disorder becomes an issue, with brothers and sisters of those with the illness being 8 to 18 times more likely to develop depression.

Research into the genetic causes of depression has attempted to find a specific gene associated with the disease, but there is little consistency as to exactly which gene is responsible. Just because a person may possess the nasty gene, doesn't guarantee they will get the disorder outright. But in December of 2004, researchers at the National Institute of Mental Health and the National Heart, Lung, and Blood Institute did find a mutant gene that they claim starves the brain of serotonin; this mutant gene is 10 times more prevalent in depressed patients than in regular patients with no depression. A research team at Duke University reported in the July 9, 2004 issue of _Science_ that they found some mice possessing a tiny, one-letter variation in the sequence of their tryptophan hydroxyls gene (Tph2) that results in 50 to 70 percent less serotonin and suggested that such a variant gene might also exist in humans, resulting in mood and anxiety disorders.

Fear may also have a genetic factor. Fear, ranging from general anxiety disorder to phobias, seems to be selectively bred into animals, as some experiments with mice have suggested. These "fearful" mice, if bred repeatedly with one another, seem to develop lines of fearful mice offspring. Many of us know that an anxious mother makes for an anxious child, but whether this is genetic or learned, or a combination of the two, is still up in the air.

Apparently, mice lacking functional nerve cell receptors for specific neurotransmitters, such as gamma-amino butyric acid (GABA), seem to be more fearful than mice with normal receptors. GABA is a calming amino acid used by the higher brain to repress anxiety impulses in the lower brain. Much of the study of human fear and anxiety involves neurotransmitters and receptors, and how well those receptors take in certain chemicals such as GABA and serotonin. But finding one gene that passes fear and anxiety down from one generation to the next has eluded scientists, and it may well be that the genetic component of fear is more complex, and involves particular combinations of genes controlling specific neurotransmitters and receptors, as well as the presence or lack of presence of certain brain chemicals necessary for well-being.

As more research into the genetic causes of depression, anxiety, panic attacks, even alcoholism and drug addiction is carried out, we will, no doubt, learn if we indeed are feeling the fear and pain of our ancestors. In his stunning and intriguing book, *Watermark: The Disaster That Changed the World and Humanity 12,000 Years Ago*, author Joseph Christy-Vitale speculates that a catastrophic event might indeed be responsible for today's growing rates of depression, generalized anxiety, and other emotional and mental disorders with which modern humans deal.

Christy-Vitale's book tells the story of Phaeton, the Shining One, a piece of a supernova that came so close to Earth that it caused massive destruction and devastation. To those who lived when Phaeton struck, approximately 12,000 years ago, the horror of watching such an event take place, and of witnessing the ensuing death and destruction, was, as the author suggests, so traumatizing that it affected the survivors on such a deep, profound level that we, their descendants, are still feeling the post-traumatic stress of that event.

The author asks, "Is it possible for something that occurred so long ago to still affect us today?" Yes, he believes. "Since Phaeton, societies and civilizations have continuously retraumatized themselves through child-rearing methods, education, social custom, racism, religion, war, predatory economic policies leading to inequity and poverty, and a separation from nature resulting in the destruction of Earth's environment." He also points to the damaging and unbalanced behaviors leading to addictions, which our current belief systems seem unable to explain or address. Christy-Vitale quotes Chellis Glendinning, who wrote *My Name Is Chellis & I'm In Recovery From Western Civilization* that "Every trauma that occurs is an individual trauma...every trauma is a social trauma with roots in social institutions and implications for society at large, and every trauma is a historic trauma, fostered by the past and reverberating into the future."

Trauma passed down from generation to generation, Christy-Vitale warns, is a recipe for a cultural disease that can become deeply institutionalized within a civilization. We, the modern descendants, may carry the genes of traumatic memory, but we may also carry the genes of hope, renewal, and courage.

Not every culture adapts negatively over the long run to such severe and all-encompassing trauma, and the survivors of Toba, no doubt, developed positive psychological adaptations in order to make sense of their new situation, and, once the environment allowed, moved forward and outward, expanding the reach of humanity to distant shores.

And to us, the descendants of those who, despite their fear and suffering, found a new reason to hope: a new sunrise, a new day.

CHAPTER 9

The Lurking Mega-Disaster

The time was between 10 and 12 years ago. The location was a watering hole in the middle of a savannah-like grassland in the northeastern section of modern-day Antelope County, Nebraska. It was a very warm, humid, sunny afternoon, and a variety of animal species were waiting their turns to drink and cool off in the water once the boisterous and domineering rhinoceroses departed to their resting places in the ocean of sub-tropical grasses, or the shade of a hackberry tree.

More than 40 species of plants and animals were found in and around the watering hole. The variety of animals waiting their turns included one- and three-toed horses, three species of camels (including one with a long giraffe-like neck), and sabre-toothed deer. Pond turtles, giant tortoises, and various types of birds ringed the shoreline, staying out of harm's way, waiting for the rhinos to leave. Lurking in the grass further back from the water hole were the scavengers waiting to prey on smaller species, sick, or young animals. Prominent among the scavengers was a species of dog considered to be one of the

189

Fig 9-1. There were more than 40 species of plants and animals in the area of this 10-million-year-old watering hole, an area that is now Ashfall Fossil Beds State Historical Park in Antelope County, Nebraska. Source of photo courtesy of University of Nebraska State Museum.

top predators prowling the territory at that time, generically referred to as "bone-crushing dogs" because of their powerful teeth and jaws, and hyena-like features. Predators noticeably absent from the landscape were humans, because they had not yet evolved.

Then, gradually, on this fateful day, huge, dark clouds began to appear in the western sky; these were not storm clouds or soot from grass fires. Unbeknownst to the animals down at the watering hole, these strange and ominous-looking clouds consisted of deadly ash from a huge volcanic eruption many hundreds of miles west, in what is now the modern-day state of Idaho. The Yellowstone hotspot was about to make an impact in a very deadly manner.

As the overhead sky darkened, gray ash began to fall like snow. Those animals out in the open didn't know that what they were breathing consisted of jagged pieces of volcanic rock and glass that have been described as "tiny little needles." Once the ash got into their lungs, it combined with naturally occurring moisture in the lungs to form a kind of cement. Birds and turtles were the first species to die as their lungs filled up with sediment; musk deer and small carnivores were next. The larger animals were also having trouble breathing. Inside their bodies, their bones were growing abnormal patches of highly porous new bone matter, especially around the lower jaw and on the shafts of major limbs, as well as their ribs. This is evidence that they were not getting enough oxygen, and is symptomatic of a disease known today as Marie's Disease.

Finally, even the largest animals had to return to the pond to drink. A herd of nearly 100 rhinos must have shuffled through the clouds of blowing ash to get to the pond. Some mothers had baby rhinos by their sides. After the death of the mother rhinos, many of the young kept trying to nurse until eventually they died as well. While most of the animals survived the actual ashfall, as they continued to breathe ash stirred up by their movements and the winds, and to graze on ash-covered grasses, their lungs filled with the deadly ash particles and they slowly suffocated.

The volcanic ash deposited in and around the watering hole continued to blow around like fresh snow. Eventually, the high ground was blown free of ash, but low-lying areas, such as the watering hole, were filled to depths of 8 feet or more. Undisturbed, except by an occasional scavenging meat-eater, the skeletons of the dead mammals, birds, and turtles were entombed in the blowing and drifting ash, and preserved in their death positions, complete with the remains of their last meals in their mouths and stomachs, and their last steps preserved in a sandstone deposit at the base of the ashfall. Paleontologists believe that because of the danger that the ash presented to later scavengers, the site remained relatively untouched for the next 10 million years. This included near misses by ice sheets advancing from the north during subsequent glacial periods. If an ice sheet had reached the watering hole site sometime in the intervening years, there would, most likely, be no intact animal remains to tell the story of their fateful encounter with the ashfall event 10 million years ago. The addition of a stratum of more erosion-resistant sandstone above the ash layer acted as caprock to preserve the ash layer and its fossil contents for the next 10 million years.

Ten million years later, on a summer day in 1971, University of Nebraska State Museum paleontologist Michael Voorhies, was walking with his wife Jane through a series of gullies on a farm in northeastern Nebraska. One of the gullies they happened to wander down had just been heavily eroded by torrential rains, and peeking out of the wall of the gully was a small piece of white bone surrounded by what looked like ash. Dr. Voorhies recognized the piece of white bone as part of the skull of a baby rhinoceros. Dr. Voorhies described his encounters by saying, "Excitedly, I brushed the ash away from the little skull, first from the oversized teeth, then farther back looking for evidence that the rest of the skeleton might be there. It was. Just as the old song has it:

'The head bone connected to the neck bone,
The neck bone connected to the backbone,
The backbone connected to the hip bone…'"

Dr. Voorhies had discovered what would eventually be recognized as one of the rarest kinds of paleontological finds: a konservat-lagerstatte. A konservat-lagerstatte is a deposit of exceptionally well-preserved fossils. Unlike most fossil deposits, which consist of scattered bones accumulated over extended periods of time, the ash bed contains mostly articulated remains with bones still joined together in their proper order. Remarkably life-like skeletons of hundreds of animals have been uncovered, and all finds have been left in place with sections open for public viewing. In 1991, the site became Ashfall Fossil Beds State Historical Park, and in May 2006 was declared a National Natural Landmark.

Fig. 9-2. Say hello to two of Ashfall Fossil Beds many celebrities: Morris, (foreground), a male rhino with large tusks and feet, and McGrew, (background) a female with smaller tusks, and although it can't be seen in this photo, the bones of an unborn calf in her pelvic cavity. The state of preservation of the skeletons at Ashfall Fossil Beds is amazing, and in addition to providing a detailed snapshot of life at this watering hole in Nebraska more than 10 million years ago, the paired associations seen throughout the Park (for example, a mother and her calf in a nursing position) convey an overpowering sense of the agony they experienced as they succumbed to the deadly volcanic ashfall from an eruption many hundreds of miles to the west in the present day state of Idaho. Image is Courtesy of The University of Nebraska State Museum; photographer Gregory Brown.

As noted earlier, the Ashfall Fossil Beds owe their existence to a huge volcanic eruption that occurred some 10 million years ago nearly 1,000 miles (1,600 kilometers) west of the ashfall site. The identification of a 10-million-year-old volcano in Idaho was a result of age dating and chemical analysis of ash from the Fossil Beds and comparison with similar information from volcanoes in the Bruneau-Jarbidge (BJ) volcanic field in southwestern Idaho. The BJ volcanic field is a system of nested calderas and collapse structures similar to, though larger than, structures of the present-day Yellowstone Volcanic Plateau. Between 10 and 12 million years ago, the BJ field and southwestern Idaho was located over the Yellowstone Hot Spot (melting anomaly). Throughout the next 10 to 12 million years, the North American tectonic plate slowly (about 1 to 2 inches per year) moved to the southwest, resulting in the present-day separation between the BJ volcanic field and the hot spot of about 300 miles. The path followed by the North American plate is part of the Snake River Plain (SRP), and is punctuated by episodic large-scale silicic volcanism formed as the plate passed over the hotspot. Volcanoes along the SRP progress in age from approximately 16.5 million years old at the extreme southeastern end of the track on the Nevada-Oregon border to present-day activity at Yellowstone National Park.

In 2002, geologists from the University of Utah reported that, based on their decade-long field and laboratory analyses of ashfall tuffs from the SRP, the Yellowstone hotspot produced 142 huge volcanic eruptions during the past 16.5 million years, at least 40 percent more than previously known. The eruptions are estimated to have generated 60 to 140 cubic miles of material, and in some cases, as much as 600 cubic miles. Of particular importance here was the finding that the occurrence rate of such eruptions slowed during the past 16.5 million years, from 32 giant eruptions per million years before 15.2 million years ago to 2.5 cataclysmic eruptions per million years during the past 8.5 million years. The decrease in the rate of volcanism is presumably due to the increasingly cold and thicker continental crust of the eastern portions of Wyoming that the seemingly "fixed" upper mantle hotspot impinges as the North American plate progresses to the southwest. Thus, the decreasing rate reflects the fact that it takes longer for the magma to break through the thicker and colder crust, resulting in fewer eruptions per unit of time.

While an accurate measure of the BJ eruption that produced the Ashfall Fossil Beds in Nebraska is unavailable, estimates range from at least 100 to as much as 250 times larger than the 1980 Mount St. Helens eruption. Assuming a bulk volume of 0.24 cubic miles for the Mount Saint Helens eruption, the above range in size implies that the volume of material ejected by the BJ eruption was 24 to 60 cubic miles, smaller than a supervolcano according to the criteria discussed in Chapter 4.

The death assemblage at Ashfall Beds serves as a vivid reminder of just one of the many hazards that humans would face in the event of a large volcanic eruption in the western United States. It is alarming that so many animals, large and small, succumbed in such a short period of time to the effects of what probably was not on the scale of a supervolcano. This raises the question of just how devastating the environmental consequences of an actual supervolcano, spewing out a volume of material in excess of 120 cubic miles, would be. In Chapter 3, we looked at the effects, sometimes catastrophic, on humans and their ecological and political environments caused by some of the largest, but still "normal," explosive eruptions during the past 3,000 years. Given the immensity of the effects caused by these eruptions, which were all smaller than supereruptions, we can only wonder how much more severe the aftermath of such a cataclysmic eruption in the western United States would be.

YELLOWSTONE AND LONG VALLEY ERUPTIONS

The Yellowstone National park region has produced three exceedingly large caldera eruptions in the past 2.1 million years. In each of these cataclysmic events, enormous volumes of magma erupted at the surface and into the atmosphere as mixtures of red-hot pumice, volcanic ash, and gas that spread as pyroclastic flows in all directions. The first of these caldera-forming eruptions 2.1 million years ago created a widespread volcanic deposit known as the Huckleberry Ridge Tuff. This titanic event, one of the five largest individual volcanic eruptions known anywhere on the Earth, formed a caldera more than 60 miles across. The volume of material ejected during this event has been estimated at 600 cubic miles, comparable in size to the cataclysmic YTT supereruption at the Toba caldera 74,000 years ago.

A smaller, but still huge, caldera eruption occurred 1.3 million years ago. This eruption formed the Henrys Fork Caldera, located in the area of Island Park, west of Yellowstone National Park, and produced another widespread volcanic deposit called the Mesa Falls Tuff. The estimated volume of this event was 70 cubic miles, comparable to the BJ eruption that resulted in the death assemblage of mammals at the Ashfall Fossil Beds in Nebraska.

The region's most-recent caldera-forming eruption 640,000 years ago, with an estimated volume of 240 cubic miles, created the 35-mile-wide, 50-mile-long Yellowstone Caldera. Pyroclastic flows from this eruption left thick volcanic deposits known as the Lava Creek Tuff. Huge volumes of volcanic ash were blasted high into the atmosphere, and deposits of this ash can still be found in places as distant from Yellowstone as Iowa, Louisiana, and California.

A recurrence interval of 600,000 to 700,000 years is often quoted for large caldera eruptions at Yellowstone, even though it is only based on two intervals and, as a result, of highly questionable statistical significance. It should be noted that the 1.3 million year event, while a very large eruption, was not a supervolcano, and casts additional doubt on the meaning of a recurrence interval at the Yellowstone hotspot. For instance, is the applicable recurrence interval for a supervolcano at Yellowstone more like 1.5 million years (the approximate time between the 2.1-million-year and the 640,000-year-old events), rather than 600,000 to 700,000 years which puts the 70 cubic mile event at 1.3 million years on an equal footing with the two supervolcanoes? It is still reasonable to speculate, however, that based on the three events at Yellowstone, a large eruption with a bulk volume greater than 60 cubic miles could occur within the next 100,000 years, or sooner.

Second only to Yellowstone in North America is the Long Valley caldera, in east-central California, just south of Mono Lake, near the Nevada state line. The biggest eruption from Long Valley occurred about 760,000 years ago, and involved the ejection of 140 to 170 cubic miles of high-silica rhyolite referred to as the Bishop Tuff—approximately 50 times the amount of magma ejected during the 1991 Mount Pinatubo eruption in the Philippines. Scientists estimate that as much as 12 cubic miles of tephra was dispersed with significant accumulations of ashfall as far east as Nebraska. This massive, short-duration eruption (approximately 10 days) resulted in the widespread deposition of the Bishop Tuff and the simultaneous 1 to 2 mile subsidence of the magma chamber roof to form the 10×20 mile oval depression of the Long Valley caldera.

The explosive caldera-forming eruptions at Yellowstone and Long Valley occurred when large volumes of "rhyolitic" magma accumulated at shallow levels in the Earth's crust, as little as 3 miles below the surface. This highly viscous (thick and sticky) magma, charged with dissolved gas, then moved upward, stressing the crust and generating earthquakes. As the magma neared the surface and pressure decreased, the expanding gas caused violent explosions. If another large caldera-forming eruption were to occur at either Yellowstone or Long Valley, its effects would be worldwide. Thick ash deposits would bury vast areas of the United States, and injection of huge volumes of volcanic gases into the atmosphere could drastically affect global climate for several years. Based on the discussions in Chapters 3 and 5, the effects of gas release, and the resulting aerosols, are expected to outweigh effects of ash fallout over regional to continental-scale areas and long time frames.

Recently, two modeling studies were carried out to estimate the climatic effects of a large stratospheric injection of sulfate aerosols, which would be expected from a supereruption. In one study, researchers used a state-of-the-art climate model with a sulfur dioxide input 100 times larger than that of the 1991 Pinatubo eruption from an explosive source at 45 degrees north latitude, appropriate for a Yellowstone eruption. The results indicate that the season of the year plays an important role in the global spread of aerosols. For instance, aerosols from an eruption in mid-latitude northern hemisphere exhibit more of a global spreading in summer months than in winter months. These results support findings from another modeling analysis that simulated the climatic response to a Toba-type supereruption. While both studies predict drastic cooling with average global temperatures plummeting by as much as 18°F, the studies also predict that the volcanic aerosols would leave the atmosphere fairly quickly, and that the climate would largely recover within a decade. Thus, while environmental consequences of such a temperature drop would be devastating on a short-term basis, it does not appear likely that a supereruption could be the primary cause of an ice age.

Support for the modeling results comes from ice cores recovered from Greenland and Antarctica. As described in Chapter 5, researchers, examining the ice cores, reported that a sulfuric acid peak caused by the Toba supereruption did not last as long as once thought. Instead, the acid peak was only observable in the ice cores for six years or fewer. The interpretation of the evidence from the ice cores agreed with the modeling predictions of a short-term impact on the climate—this is the good news.

The bad news comes from recent investigations of the long-term effects of sulfur dioxide injected into the stratosphere during supereruptions, compared to Tambora and smaller explosive eruptions. For sulfur dioxide to become sulfuric acid, it must be oxidized; in other words, it must acquire two oxygen atoms from other compounds already present in the atmosphere. In the case where stratospheric ozone is the interacting compound, it transfers its oxygen isotopic signature to the resulting acid. The characteristic signature is acquired as a result of rare chemical transformations that ozone gas undergoes in the presence of intense solar radiation bombarding the stratosphere. In 2003, scientists at Caltech in Pasadena, California, reported that oxygen isotopic signatures found in sulfate from Yellowstone and Long Valley ash samples implied that significant amounts of stratospheric ozone were used up in reactions with gas from the supereruptions in those regions. Similar results for other supereruptions were obtained from acid layers in Antarctica ice cores. The result of these investigations was that supereruptions could probably eat holes in the ozone layer for an even longer period of time than they take to cool the climate. This effect would not be expected from smaller eruptions because sulfur-gases emitted by these events are predominantly oxidized in the troposphere and have their principal effect on stratospheric ozone through interaction with chlorofluorocarbons (CFCs) of anthropogenic origin.

Stratospheric ozone provides a protective shield from potentially damaging doses of ultraviolet-B radiation (UV-B), and significant depletion of stratospheric ozone could lead to significant increases in UV-B reaching the Earth's surface. This would result in a wide range of potentially damaging human and animal health effects, primarily related to the skin, eyes, and immune system. Thus, while the news is good in terms of possible long-term volcanic winter type effects (that is, a possible ice age), the combination of several years of drastically reduced temperatures and even longer increased UV-B radiation levels would wreak havoc on humans and the environment. This deadly combination may well be responsible for the human bottleneck hypothesized to have occurred after the Toba supereruption 74,000 years ago.

In a report published in April 2005, the U.S. Geological Survey performed an assessment of the relative hazard of 169 U.S. volcanoes. The Long Valley and Yellowstone calderas were identified as very high and high-threat volcanoes, respectively. The threat assessments are based in large part on the present state of unrest at the calderas. Unrest includes the level of occurrence of earthquake swarms, episodes of uplift and subsidence, and heat and gas emissions.

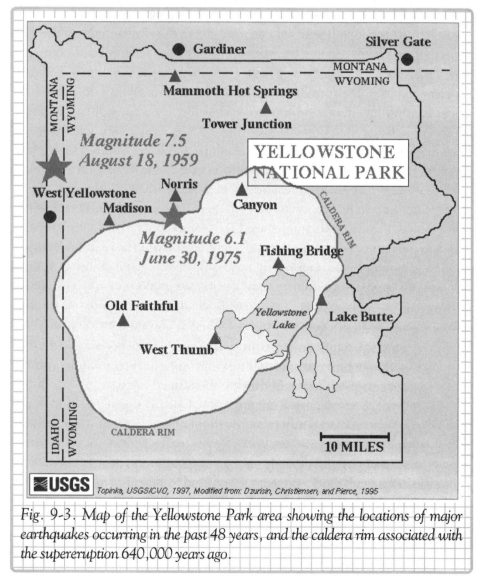

Fig. 9-3. Map of the Yellowstone Park area showing the locations of major earthquakes occurring in the past 48 years, and the caldera rim associated with the supereruption 640,000 years ago.

UNREST AT YELLOWSTONE

In the case of Yellowstone, there are typically 1,000 to 3,000 earthquakes that occur each year within the park and its immediate surroundings. Although most are too small to be felt, these quakes, recorded by seismograph systems located in and near the Park, reflect the fact that the Yellowstone caldera is an active volcano, and one of the most seismically active areas in the United States. Each year, several quakes of magnitude 3 to 4 are felt by people in the park. Although some quakes are caused by rising magma and hot-ground-water movement, many emanate from regional faults related to crustal stretching and mountain building. The most notable earthquake in Yellowstone's recent history occurred in 1959. Centered near Hebgen Lake, just west of the park, the earthquake had a magnitude of 7.5 and killed 28 people, most of them in a landslide that was triggered by the quake. Geologists believe that large earthquakes such as the Hebgen Lake event are unlikely within the Yellowstone Caldera itself, because subsurface temperatures there are high, weakening the bedrock and making it less able to build up sufficient strain to result in a rupture. However, quakes within the caldera can be as large as magnitude 6.5. A quake of about this size occurred in 1975 near Norris Geyser Basin, and was felt throughout the region.

As for deformation of the caldera, geologists most familiar with the Yellowstone region point out that while the caldera has been relatively dormant in terms of major volcanic activity for more than 70,000 years, it has been rising and falling for at least the past 15,000 years, at times more than 10 feet. In more recent times, between 1923 and 1975, the entire Yellowstone caldera rose about 3 to 4 feet. Suddenly in 1985, it reversed and started subsiding; in 1990 it reversed again and started to inflate.

Between 1997 and 2003, the northern part of the Yellowstone caldera began to bulge. The bulge, measuring about 25 miles across, rose approximately 5 inches. Simultaneously, there was a sudden rise in temperatures, new steam vents, and the awakening of the geysers in the area. Steamboat Geyser, dormant for nine years, erupted in May 2000, and then several times between 2002 and 2003. The nearby Porkchop Geyser awoke after 14 years of dormancy. Unusual thermal phenomena at the nearby Norris Geyser Basin resulted in such high ground temperatures in 2003 that Yellowstone officials decided to close some boardwalks out of fear that visitors might be burned.

UNREST AT LONG VALLEY

We now turn to the Long Valley caldera. In 1978, a Richter magnitude (M) 5.4 earthquake struck 6 miles southeast of the caldera. This event ended two decades of relatively low earthquake activity in the area, and was followed in May 1980 by an earthquake swarm that included four strong M 6 earthquakes. These events struck the Mammoth Lakes area on the southern margin of the Long Valley caldera. The largest of the M 6 earthquakes occurred one week after the Mount St. Helens eruption of May 18. A report in the *Mammoth Times* newspaper dated September 7, 2000, recalled how the southern California news media in late May 1980, through implicit association of the Mount St. Helens eruption and the strong earthquakes near Mammoth Lakes, had Mammoth Mountain ready to blow. The *Mammoth Times* report went on to note that, "local governments worked to create an all-encompassing emergency plan. Mammoth got an escape route, known euphemistically as the Scenic Loop."

In response to continuing seismic activity and uplift observed within the central portion of the Long Valley caldera, the U.S. Geological Survey issued a "Notice of Potential Volcanic Hazard" in 1982. The notice was blamed for causing a severe drop in tourism, and also a downturn in what had been a boom in the Mammoth Lakes housing market. When, after a period of time, a volcanic eruption didn't happen, the residents of Mammoth Lakes began referring to the USGS as the "U.S. Guessing Society," among other names.

An ominous sign of unrest beneath the Long Valley caldera occurred in 1989 when magma intruded beneath Mammoth Mountain, a stratovolcano located on the western rim of the caldera, and the site of a major ski resort. While magma did not erupt, large volumes of carbon dioxide were discovered seeping into the soil in an area known as Horseshoe Lake. In a paper published in *Nature* in 2002, scientists reported the results of a soil-gas survey begun in 1994 where they observed carbon dioxide concentrations of 30 to 96 percent in a 75-acre region of dead trees. Based on their study, they concluded that although the tree kill coincided with the episode of shallow dike (magma) intrusion, the magnitude and duration of the carbon dioxide flux indicated that a larger, deeper magma source and/or a large reservoir of high-pressure gas was being tapped.

In 2003, researchers reported the results of geodetic and gravity surveys conducted in the Long Valley caldera. They concluded that the results of their surveys did not support hydrothermal fluid intrusion as the primary cause of unrest, instead confirmed the intrusion of silicic magma beneath the caldera. Various signs of unrest continue at the present time, and should serve as a reminder of the boiling cauldron that lurks below the surface.

EARTHQUAKES AS TRIGGERS?

On November 3, 2002, the largest inland earthquake in North America in almost 150 years struck Alaska, about 85 miles south of Fairbanks. This M 7.9 event, known as the Denali fault earthquake, ranks among the largest strike-slip ruptures of the past two centuries. Its length and slip magnitudes are comparable with those of the great California earthquakes of 1857 and 1906. Horizontal displacements of up to 26 feet were measured along sections of the roughly 200-mile-long fault. Large-amplitude surface waves from the Denali Fault earthquake triggered a series of earthquake swarms and changes in geyser activity at Yellowstone, and strain offsets and microseismicity under Mammoth Mountain on the rim of Long Valley caldera. The amazing thing is that the Denali fault earthquake was about 1,940 and 2,160 miles distant from Yellowstone and Long Valley, respectively. This leads to the question of what impact a very large (M > 7.75), but closer, earthquake might have on the calderas. For instance, could a large earthquake in southern California trigger an eruption at either Long Valley or Yellowstone? Major earthquakes in southern California in 1992 and 1999 suggest a possible answer to this question.

On June 28, 1992, an M 7.3 earthquake occurred near the town of Landers, approximately 100 miles east of Los Angeles in the Mojave Desert of southern California. Even though the earthquake was more than 770 miles southwest of Yellowstone, the large-amplitude surface waves from this event changed the relatively periodic eruption rate of one of the geysers in the park from approximately 56 minutes to an erratic rate for about 34 hours, and triggered a swarm of small earthquakes beneath the caldera.

In their *Summary of Long Valley Caldera Activity for 1992*, the USGS scientists commented:

Certainly the most remarkable and energetic event in the caldera during 1992 was the abrupt surge in local seismicity that began immediately following the June 28, M=7.3 Landers earthquake, which was located in southern California some 400 km south of the caldera. The surge in seismicity triggered by the Landers earthquake was the strongest "swarm" in the caldera during 1992. It included two M>3 earthquakes and over 250 smaller events in the first six days after the Landers earthquake.

Signals associated with transient periods of deformation near the western boundary, and within the caldera, also accompanied the triggered seismicity.

In the October 29, 1998, issue of *Nature*, two scientists from the Carnegie Institution of Washington reported results of an analysis of the historical record of earthquakes and volcanic eruptions conducted to see if there are significantly more eruptions immediately following large earthquakes. They noted that their study was, in part, motivated by the triggering of seismicity and deformation at the Long Valley caldera by the 1992 Landers earthquake. They found that, "within a day or two of large earthquakes there are many more eruptions within a range of 460 miles than would otherwise be expected. Additionally, it is well known that volcanoes separated by hundreds of kilometers frequently erupt in unison; the characteristics of such eruption pairs are also consistent with the hypothesis that the second eruption is triggered by earthquakes associated with the first."

On October 16, 1999, a year after the *Nature* paper appeared, another major earthquake occurred in the Mojave Desert in southern California. This event, referred to as the Hector Mine earthquake, was slightly smaller than the Landers earthquake, weighing in at M 7.1. The epicentral distances from the Hector Mine earthquake to the Yellowstone and Long Valley calderas were comparable to distances from the Landers event. While we are not aware of any reported triggering of activity at Yellowstone from the Hector Mine earthquake, deformation associated with the earthquake was observed at Long Valley, and was followed 20 to 30 minutes later by a swarm of small earthquakes localized beneath the north flank of Mammoth Mountain.

While the evidence for remote triggering of unrest at volcanic calderas by large earthquakes is reasonably convincing, what's the possibility for an earthquake occurring in southern California with a significantly larger magnitude than the 1992 Landers event? Two studies that are relevant to this question were reported in 2005 and 2006. These studies consider the potential for a large earthquake along the southern section of the San Andreas Fault, which has been ominously quiet for way too long.

THE SAN ANDREAS FAULT

The San Andreas Fault is a continental transform fault that accommodates the relative right-lateral motion of the Pacific and North American tectonic plates. During a major strike-slip earthquake a person standing on the western, or Pacific, side of the fault would see someone on the eastern, or North American, side move to the right. On June 22, 2006, Yuri Fialko, an associate professor of geophysics at the Institute of Geophysics and Planetary Physics, Scripps Institution of Oceanography, University of California at San Diego in La Jolla, California, reported in the journal *Nature* that the southernmost segment of the San Andreas fault running from San Bernardino in the north to the Mexican border in the south is primed for a major earthquake with between 20 and 30 feet of displacement possible, most likely in a right-lateral strike-slip sense. This section of the San Andreas has not experienced a major earthquake for more than 300 years. The last major earthquake, an M 7.7, occurred in 1690 and ruptured about a 140-mile-long segment of the fault. Although a major event, it went largely unnoticed because hardly anyone lived there at the time. Professor Fialko's study, which was based on both satellite and ground-based measurements of large-scale deformation, indicates that stress has been building up since then, and that this southern segment of the fault may be approaching the end of the interseismic phase of the earthquake cycle. He is quoted by the Scripps Howard News Service as saying, "All these data suggest that the fault is ready for the next big earthquake, but exactly when the triggering will happen and when the earthquake will occur we cannot tell. It could be tomorrow or it could be 10 years or more from now." An earthquake in southern California would only be, on average, about 280 and 770 miles from the Long Valley and Yellowstone calderas, respectively, and could possibly trigger significant activity at either, or both, of those sites.

In the May 13, 2005 issue of *Science*, four scientists reported results of an investigation of spatial and temporal characteristics of large earthquakes that occurred since A.D. 1200 along the southern 340 miles of the San Andreas

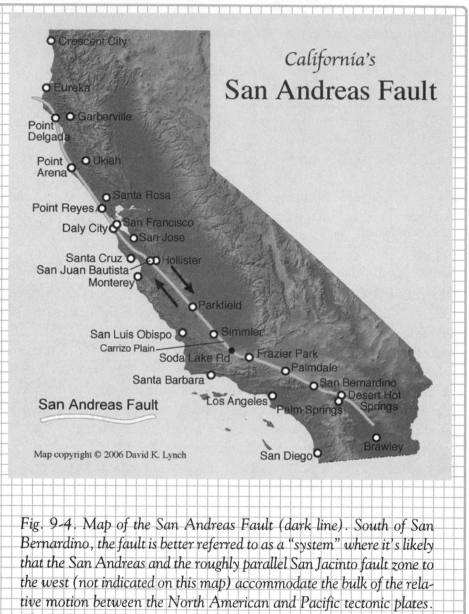

Fig. 9-4. Map of the San Andreas Fault (dark line). South of San Bernardino, the fault is better referred to as a "system" where it's likely that the San Andreas and the roughly parallel San Jacinto fault zone to the west (not indicated on this map) accommodate the bulk of the relative motion between the North American and Pacific tectonic plates. The 1857 M 8 Fort Tejon earthquake referred to in the text is thought to have originated near Parkfield in the north and extended southeast for about 220 miles (360 km) to the San Bernardino area. Image courtesy of David K. Lynch, "Field Guide to the San Andreas Fault" Thule Scientific, Topanga, California.

Fault system. This section of the fault includes the 1857 M 8 Fort Tejon earthquake, which ruptured between 180 and 220 miles of the northern portion of the section studied. They identified geologic records for up to 56 earthquakes, although some of the records may pertain to the same event. The lead author of the *Science* paper noted that most of the fault ruptures every 200 years, but because of uncertainty in dating the individual ruptures, they were not able to tell whether it was one earthquake or a number of closely timed earthquakes. The scientists determined the probability of the current lull in activity ending in the next 30 years to be 20 percent, 40 percent, and 70 percent, depending on how the earthquake ruptures were modeled in space and time. The principal conclusion of their study was that the next rupture may be one great earthquake of M8 or greater, or a series of large earthquakes all smaller than M 8. In either case, the southern section of the San Andreas is locked and loaded.

The lack of experience that scientists have when it comes to recognizing the signs or signals coming from a volcanic system or caldera on the verge of a supereruption poses a major challenge to successful forecasting. Recall that the last supereruption, Taupo, occurred about 26,500 years ago. In addition, the last VEI 7 eruption, Tambora, occurred in 1815, well before instrumental records became available. Thus, volcanologists will be hard pressed to not only identify a signal from an eruption that could be imminent, but recognize a bona fide signal (for example, volcanic tremor) for a VEI 7 or 8 eruption, as opposed to a 4 or 5.

In a paper published in 2006 in the *Philosophical Transactions of the Royal Society*, scientists from the USGS and the University of Utah made the following concluding remarks:

> With limited experience monitoring and responding to large-scale volcanic crises, society cannot expect a 100-percent success rate at avoiding future volcanic catastrophes. We can, however, make sure that we learn from the next VEI 6 or 7 eruption, by recording a full spectrum of signals emitted prior to eruption. At present, only a small fraction of Earth's high-threat volcanoes is monitored in a manner that would provide a useful history of the run-up to a volcanic disaster. If we are to reduce the risk from future large eruptions, we will need to do better.

CHAPTER 10

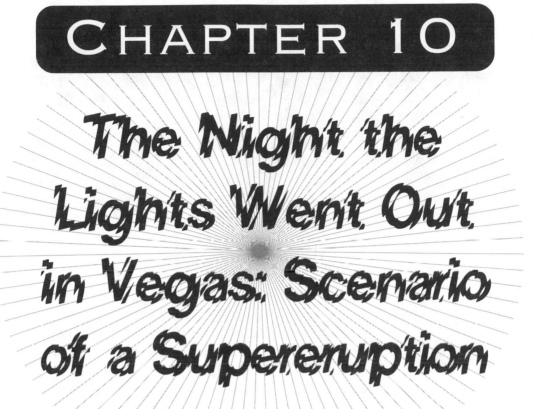

The Night the Lights Went Out in Vegas: Scenario of a Supereruption

The explosion will be heard around the world. The sky will darken, black rain will fall, and the Earth will be plunged into the equivalent of a nuclear winter.
—From *Supervolcanoes, BBC.co.uk*

AUTHORS' NOTE

The following chapter presents a fictional scenario of a supereruption. Although this is a speculative idea of what might happen should a supervolcano erupt in the near future, in this case the year 2015, it is based upon scientific observations and the most cutting edge research we have available at this time.

Although much media fuss has been made in recent years about the potential for a supereruption at Yellowstone, thanks to a slew of movies, documentaries, and even books, and while we the authors do agree that Yellowstone could be next in terms of a supervolcanic event, we have chosen to focus instead on a much more probable event, especially in light of new evidence linking potentially massive earthquake activity along the southern section of the San Andreas Fault to the Long Valley Caldera in California. Recent scientific research suggests that a great earthquake on this fault could be imminent. We believe it could possibly trigger something far more devastating.

But ultimately, whether it is Yellowstone that next erupts, or Long Valley, or another mighty supervolcano that exists somewhere in the world that is not even on the radar screen, it will, indeed, change the course of human history once again. Life as we know it will cease to exist, and a new idea of civilization will be forced to take shape out of the ashes.

In his book *Apocalypse*, Bill McGuire makes the following observation concerning the magnitude of possible future eruptions at the Tavurvur volcano in Papua New Guinea, and at the Campi Flegrei, Italy, and Long Valley calderas.

While Tavurvur rumbles on, and Campi Flegrei and Long Valley may erupt soon, it is highly unlikely that any of these calderas will host the next super-eruption. The short-term swelling that contributes to the restless state of these calderas is simply not persistent enough, nor sufficiently large and extensive, to be attributable to the volume of fresh magma required to feed a super-eruption. The unfortunate truth is that the next super-eruption will probably occur at a poorly monitored or unmonitored volcano that we know little about, located in an obscure, perhaps even largely uninhabited, part of the world, or even from a site where no currently-active volcano or restless caldera field exists.

Under normal circumstances we would be inclined to agree with McGuire's assessment, especially concerning the possibility of a supereruption in the Long Valley caldera.

But let's jump ahead to the year 2011, when circumstances in central and southern California are anything but normal!

SCENARIO OF A SUPERERUPTION

In central California, the Long Valley caldera began to show signs of increasing unrest. In March 2011, a swarm of earthquakes, unlike the 1980 swarm, which struck the southern margin of the caldera, occurred near the eastern rim of the caldera north of Lake Crowley. The swarm included several earthquakes with magnitudes in the M 5.5 to 6.0 range. Simultaneously, the dome-like uplift in the center of the caldera, which had risen nearly 23 feet since the episode began in 1980, was on the rise again. This time, however, the rate of uplift was faster, and the area affected larger than corresponding conditions prevailing at the turn of the century. New microgravity surveys indicated the presence of massive amounts of low-density, rhyolitic, or very sticky, magma over and above the calculated amounts observed in 2003.

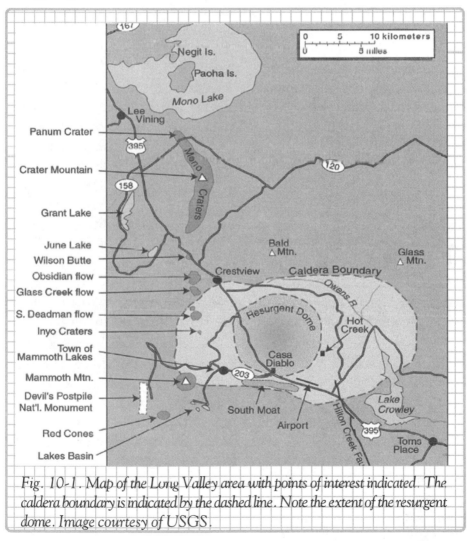

Fig. 10-1. Map of the Long Valley area with points of interest indicated. The caldera boundary is indicated by the dashed line. Note the extent of the resurgent dome. Image courtesy of USGS.

Another sign of unrest, seepage of carbon dioxide gas through the soil, was now detected near the eastern rim of the caldera at concentrations between 60 to 95 percent of the gas content of the soil; soil gas normally contains 1 percent or less of carbon dioxide. These concentrations are similar to levels measured in the Mammoth Mountain—Horseshoe Lake tree-kill areas starting in the late 1980s. Scientists think that the ongoing episode of high carbon dioxide emission, about 1,300 tons per day in the Mammoth Mountain area, is the first large-scale release of the gas on the mountain in at least 250 years, because the oldest trees in the active tree-kill areas are about that age.

In response to this escalating unrest, the USGS, in cooperation with the California Office of Emergency Services and local authorities, issued a public notice that the condition color (level) of the Long Valley caldera was being raised from green (no immediate risk) to yellow (watch). Only one condition color/level remained before the alert of an ongoing eruption would be issued. Because the recent activity was relatively far from any sizeable population centers, such as Mammoth Lakes and the ski resort areas, there was not a lot of interest, and certainly no panic, generated as a result of the USGS announcement. In fact, most people simply ignored the warning...at their own risk.

Meanwhile, at the federal level, both FEMA and Homeland Security were put on standby, with constant updates given to the undersecretary of FEMA by the USGS scientists in charge of monitoring at the Long Valley Volcanic Observatory. USGS had developed a top-notch early warning system, and all over the country, universities and research centers worked in combination with local, state, and national offices of emergency services to keep information flowing. But for now, most of the interest, and anxiety, was focused on the shaking ground, as more quakes continued to rock and roll the area.

At the same time that Long Valley was showing signs of heightened activity, earthquake activity in the magnitude interval of M 5 to 6.5 began to occur on faults in the vicinity of the southern section of the San Andreas Fault zone in the Brawley and Salton Sea areas. Recall from Chapter 9 that this is the southern end of the section of the San Andreas that last ruptured in 1690 with an M 7.7 earthquake, and is considered locked and loaded for the next great event. The current M 5 to 6.5 seismic activity was spread out enough so that no warnings were issued by any of the responsible agencies.

The heightened activity continued at Long Valley and in southern California for four more years, creating a state of heightened suspense at USGS, and the national offices of FEMA and Homeland Security, all of whom were on standby, waiting for any sign of a major change. That is, until Tuesday, January 6, 2015, at 2:05 a.m., when the southern portion of the San Andreas Fault unlocked and ruptured in a great M 8.1 earthquake, the Big One feared by millions for decades. The rupture started near the southern end of the Salton Sea east of Brawley, California, and propagated to the north a distance of about 190 miles at a velocity of about 1.5 miles per second, finally stopping about 30 miles north of Wrightwood, California, approximately two minutes later. For many people, it was the longest 2 minutes of their lives.

The maximum displacements were later determined to be 20 to 30 feet in a right-lateral strike-slip sense. Important characteristics of the earthquake that played a crucial role in determining its impact on the level of unrest in the Long Valley caldera were that this was a shallow event, in fact, a surface rupture, and that the direction of rupture propagation, referred to as rupture directivity, was in the general direction of Long Valley itself. These properties of the rupture meant that the amplitude of the ground motion produced by the surface waves from this great earthquake, when they reached Long Valley within a few minutes after the origin time of the event, were larger than anything the caldera had experienced since the relatively close Owens Valley earthquake of 1872 with a magnitude between 7.6 and 8.

The difference now was that the caldera was primed for action.

The basic types of seismic waves generated when a fault such as the San Andreas ruptures beneath the Earth's surface include, in their order of arrival at a distant location: a P or primary wave, an S or shear wave, and two types of surface waves. It is the surface waves that produce the largest ground motions at distant locations. In the previous chapter, we saw how surface waves, known as Rayleigh waves, from the 2002 Denali Fault earthquake in Alaska, and the M 7.3 1992 Landers and the M 7.1 1999 Hector Mine earthquakes, both in the Mojave Desert area of southern California, triggered strain deformations and seismic activity at Mammoth Mountain on the south rim of the Long Valley caldera. In a similar fashion, when the Rayleigh waves from the M 8.1 southern California earthquake arrived at the Long Valley caldera, the level of earthquake activity there did two things: First, seismicity increased dramatically, and second the activity started to occur within portions of the caldera that had been seismically quiet for several decades and, in a particularly ominous sign, along most of the caldera rim.

At the same time, carbon dioxide emissions increased in the previously active areas and started in new areas. And reports of "rotten egg" smells started coming in from different locations around the caldera boundary. All these signs of increasing unrest would continue for the next several days, but in the absence of any eruptions, attention remained focused on southern California and the search and rescue operations in the wake of the earthquake.

The 8.1 magnitude of the southern California earthquake, a full magnitude unit larger than the 1999 Hector Mine earthquake, resulted in surface wave amplitudes at the Long Valley caldera 10 times larger than the 1999 earthquake. However, the barrage at the Long Valley caldera didn't cease after the seismic ground motion from the great earthquake died down. Earthquakes around the world, especially shallow events, are followed by sequences of smaller earthquakes referred to as aftershocks. In the case of a great earthquake, similar to the M 8.1 event of January 6, the aftershock sequence can last for more than one or two years. For instance, there are reports from occupants of Fort Tejon that aftershocks were felt there on a weekly basis for more than a year after the 1857 great M 8 earthquake. The 1857 earthquake was named the Fort Tejon earthquake because the fort had the dubious honor of being located very close to the mid-point of the 220-mile-long rupture zone.

The southern California earthquake resulted in hundreds of highway-related fatalities as freeway bridges collapsed and roads crumpled. The occurrence of the main shock and the large aftershocks in the early morning hours, long before rush-hour traffic would have started, however, kept the death toll down. While the Los Angeles basin shook like the proverbial bowl of jelly, as predicted by a study funded by the National Science Foundation way back in the mid-1990s, the long period ground motions associated with the earthquake did not result in significant damage to most of the family dwellings. Chimneys crumbled and carports caved in. Windows shattered, along with nerves. But most people felt they came out of the Big One intact and lucky. Years of retrofitting, and a growing awareness of the necessity for earthquake preparedness, especially after several magnitude 8 quakes hit Japan and Indonesia in the years before 2015, had given many Los Angelinos the edge on the Big One.

High-rise commercial buildings in downtown Los Angeles that were more than 10 stories high underwent some big time gyrations, but again, given the early hour, most of these buildings were unoccupied, and the number of reported accidents due to falling objects inside and outside of buildings was minimal. There was major clean up to be done, inside buildings and on the streets outside, but few lives were lost, and many of those turned out to be heart attack victims who simply gave in to the stress of the event.

Strong shaking was also felt throughout southwestern Nevada, particularly in the sediment-filled Las Vegas valley. The newly built high-rise casinos rocked and rolled—so much so that some people were almost enticed to abandon their Double Diamond slot machines. Almost...

Closer to the fault, the situation was quite different. Cities from Wrightwood and San Bernardino in the north to Brawley and El Centro in the south had declared states of emergency, and were faced with the daunting task of tending to the injured, and identifying and disposing of the dead. The fatalities numbered in the thousands, and the total rose as the days went on, with the Red Cross coming in and asking for help from the National Guard to contain the area and stop ongoing looting. The task was made extremely difficult because most of the roads and highways serving these communities were in shambles. Many mountain and high desert communities were left stranded by road closures due to rock and snow slides, and two of the most important highways for egress from the inland areas, Interstates 10 and 15, were rendered impassable in several locations. In particular, Highway 15 through the Cajon Pass was especially hard hit, and it would be weeks before all lanes, north and south, were open. What the people didn't know, however, was that if they had any hope of getting out of the area at all, they didn't have weeks, they only had five days.

Something much bigger than the Big One was brewing underground not far away.

Aftershocks continued to pound the Long Valley caldera, as well as all of southern California, in the days following the M 8.1 mainshock. While the magnitude of the aftershocks decreased fairly rapidly, strong earthquakes, with magnitudes as large as M 7 to 7.2, continued to occur during the days immediately after the mainshock. The situation between the aftershock sequence and the caldera is analogous to a boxing match where one boxer is the aggressor (the aftershocks) forcing the other boxer into the role of a defender (the caldera). The aggressor pounds away with a flurry of jabs, punctuated by an occasional whopper of a punch, to the head and stomach of the defender.

Gradually, the aggressor wears his opponent down to the point where one well-aimed punch puts the defender down on the mat for the 10-count.

By January 8, the aftershocks had the caldera on the ropes. On that day, seismograph systems deployed throughout the Long Valley area began to record the dreaded signature of "harmonic tremor." Magma was on the move. Then on January 9, the harmonic tremor ceased, and it was quiet for about two hours before an M 6.3 earthquake struck in the Lake Crowley area on the southeast corner of the caldera. This was the largest earthquake in the Long Valley area since the M 6 earthquakes back in 1980. Small aftershocks were felt throughout the rest of that day and into the next, January 10, when around 2:30 p.m. another major earthquake, this time with an M 6.4, occurred near the northeastern rim of the caldera. At this time, the USGS, again in cooperation with the California Office of Emergency Services and local authorities, issued a public notice that the condition color (level) of the Long Valley caldera was being raised from yellow (watch) to orange (warning). FEMA and Homeland Security stepped up their own actions, beginning a contingency plan that had been put into place shortly after the devastating Category 3 Hurricane Katrina of 2005 had blatantly exposed the nation's inability to cope with a major disaster. But Katrina was nothing compared to what was to come.

Many people in the area heeded the warnings that were broadcast repeatedly over radio, television, and the Internet, even major cell-phone service carriers were providing hourly updates. Masses of evacuees packed up quickly and left for points north and northeast, given the state of affairs in southern California and southwestern Nevada left by the earthquake.

And then the unthinkable happened.

It started on January 11th around 1 p.m. As millions of people went about their business on a clear, cool day, a vent erupted along the northeastern rim of the caldera with a VEI of 4. This was smaller than the VEI 5 1980 Mount St. Helens eruption, but still considered a relatively violent eruption. A moderate pyroclastic flow raced out of the vent at speeds in excess of 62 mph and covered the vicinity around the eruption out to a distance between 3 and 6 miles, barely sparing the local airport. Amazingly, no one was caught in the flow, and those few who hadn't heeded the initial warning announced by the USGS decided to leave as quickly as possible and stay away until the authorities announced it was safe to return. As it would turn out, it would be a year or more before people could even think about returning to the immediate Long Valley area.

With both the low and high altitude winds blowing out of the northwest, a cocktail of volcanic gases and tephra from the eruption formed an eruption cloud that headed toward southern Nevada and Arizona. The larger tephra fragments, bombs, and blocks, were forced out of the eruption cloud relatively close to the vent, while the lighter gases and ash spread over much larger distances. The condition level was now raised to red (eruption in progress) Level 4, corresponding to a strong explosive eruption. Everyone anywhere near the eruption was hoping and praying that this eruption would be the first and last. Unfortunately, their hopes and prayers would not be answered.

Four hours later, following another earthquake greater than M 6, a second vent started to erupt and would continue to erupt for more than 28 hours. This vent, much larger than the earlier eruption, was located in the southeast corner of the caldera close to Lake Crowley, its pyroclastic flow devastated everything left after the first eruption out to a distance of 12 to 15 miles. Fortunately, because of the earlier evacuation there were no known fatalities as a result of either eruption.

The VEI was in the mid-5 range, larger than the Mount St. Helens eruption, and a strong vertically directed column of tephra and gases reached into the stratosphere in less than 20 to 30 minutes. With the prevailing winds out of the northwest, a characteristic mushroom-shaped ash cloud formed several miles downwind from Long Valley. Moving at an average speed of close to 31 miles per hour, the cloud reached Las Vegas in less than four hours, and Phoenix about an hour after that. The cloud continued on and reached Salt Lake City in about seven hours, and Denver 12 hours after the initial eruption.

The media was now in high gear. The burning question that they asked over and over was: "Will there be a supereruption like the one 760,000 years ago?" Of course, the only honest answer at this point by scientist or prophet alike was: "We don't know." That response prompted the next question: "Is a supereruption possible?" And the much-anticipated response: "Yes." This line of questioning aimed at Earth scientists is really unfair, and drives home the fact that very often the media is more interested in a quick response that will titillate their audience rather than what they, the media, consider to be the boring, mundane truth. What the world was about to learn the hard way was that the quote by Will Durant that lead off Chapter 5, as much as any scientific conjecture, hypothesis, or theory, speaks to the heart of the matter: "Civilization exists by geological consent, subject to change without notice."

The Governor of California would issue an immediate Major Disaster Declaration, and she would be joined by the governors of the most affected states. Nevada, Colorado, Arizona, New Mexico, Wyoming, Utah, Nebraska, and Kansas would be next, followed by Idaho, Texas, Oklahoma, and South Dakota.

FEMA and Homeland Security would press the president for a Presidential Major Disaster Declaration. The following day, Madame President would do exactly that, broadcasting the declaration in an emergency news conference that would be watched by more people than all the combined episodes of American Idol, now in its final season due to lack of talent. And yet, things would only continue to worsen. The president would order all federal buildings to use recycled air, and would demand the shutting down of all air conditioners nationwide to avoid as many ash-related fatalities as possible. Some would listen; others would never hear the warning, turning on their air conditioners to clear the stale air indoors and losing their lives because of their own ignorance.

Hospitals and trauma centers, what few remained after the privatization of the hospital industry shut down many struggling facilities in lower income areas, would be overrun with people blinded by ash particles, coughing up blood as their lungs fought for clear, clean air. Injuries from roofs collapsing under the weight of ash, made even worse in areas with rain, would increase dramatically, rendering some trauma centers unable to handle the load, and forced to close their doors to additional incoming emergencies.

Throughout the next five days, eruptions originated at ring fractures located in different locations of the caldera boundary. While it couldn't be known at this time due to the lack of direct observations, the eruptions were circumscribing the boundary of the caldera formed 760,000 years ago. This would surprise most scientists when it became known at a later time. The feeling had been that the magma reservoir beneath the caldera wasn't sufficiently large enough to generate something on the scale of a supereruption. Scientists at Rensselaer Polytechnic Institute, however, may have supplied the explanation in the March 2007 issue of the *Geology* journal. Basically, they determined that there was a massive injection of hot magma underneath the surface of the Long Valley caldera sometime within 100 years of the supereruption 760,000 years ago. Their findings indicated that the introduction of this hot melt led to the massive eruption that formed the caldera. The unrest at Long Valley dates back at least to 1978, but could have started earlier and gone unnoticed given the lack of close-in monitoring instrumentation, such as seismograph systems, prior to the late 1970s. Thus, the unrest that was observed for approximately the past 50 years, and which might have started much earlier, could have been a sign of the introduction of large amounts of magma into the chamber beneath the caldera.

By January 16, the sixth day of eruptions, ash had spread as far as Nebraska to the east and into Mexico south of San Diego as upper level winds continued to shift to more northerly directions. People living in the mountain communities and inland areas of southern California were unable to flee from the advancing and descending ash cloud because of the road closures, and the highways and bridges destroyed during the great earthquake on January 6. All they could do was stay inside their homes and listen to battery-powered radios for emergency instructions. Their three-day emergency rations gone, they would begin to panic.

The emergency broadcast specifically warned against trying to operate vehicles in the midst of the choking ashfall. The biggest threat during these first few days was to the water supply. While people in isolated communities still had tap water, they found that ash had mixed in the water supply. Some of them didn't realize that they could let the water sit in a large container and the ash would settle to the bottom leaving clear water that could be scooped out. Instead, they ventured out in the ashfall in an attempt to find sources of clean water. While they wore homemade masks to protect their lungs, they did not have any extra protection for their eyes, and quickly realized just how abrasive volcanic ash. Exposure for too long a period could result in permanent eye damage and significant loss of vision. Better to stay at home, seal all windows and doors and hope for arrival of emergency services, if those services had the means for getting into the most devastated areas. And right now, not even helicopters could venture close enough to drop supplies. The thick ash made that all but impossible, at least for the duration of the eruption.

However, the month of January is about the mid-point of the rainy season in the southwest, and Mother Nature was cooking up another whammy.

Ordinarily, most people would consider rain a welcome visitor. Most, that is, except for those who choose to live on the sides of steep canyons in areas susceptible to landslides. A not-uncommon occurrence, especially along the southern California coast and coastal mountain ranges, is the formation of what is known as a cutoff low-pressure area. This is a closed low-pressure system that breaks away from the prevailing high altitude steering winds, typically the jet stream, and can linger for days, producing considerable amounts of precipitation in the southwest. The ash that had accumulated during the first two to five days, depending on the particular location relative to the prevailing winds, was a dry ash. The problem is that, once the rains came and soaked what had been dry, uncompacted ash, the now wet, compacted ash exerted loads on roofs throughout the southwest that could not hold up. This was true for many older structures and people who thought they were safe in their homes soon found themselves faced with a new hazard: collapsing roofs. Injuries, and in some cases death, began to mount as the combination of ash and rain continued to fall.

Massive explosive eruptions continued for five more days, with the grand finale coming on January 21. Exsolving gases expanding in the magma made a much larger volume of frothy fluid. The expanding, low-density, hot gases and magma mixture rose rapidly along the ring fractures, and vented at the surface as sustained explosions of white-hot froth. Driven by the hot gases, giant fountains of incandescent ash at temperatures near 1,800°F burst from the ring fractures. Plumes of ash were jettisoned into the stratosphere where planetary winds carried them around the world, eventually blanketing hundreds of thousands of square miles with volcanic ash. Nearer to the vents, fiery clouds of dense ash, fluidized by the expanding gas, boiled over crater rims and rushed across the countryside at speeds more than 60 mph, vaporizing plant and animal life in its path out to distances well beyond the limits of the earlier eruptions that started on January 11. Gaping ring fractures that extended downward into the upper magma chamber provided conduits for continuing foaming ash flows. Fractured pieces of rock comprising the volcanic edifice plunged down into the chamber, forcing additional magma up the outside edges of the ring. The collapse of the chamber roof occurred as the result of a single massive eruption, which ejected the largest amount of tephra so far in the sequence, and generated sound waves to let people, who were within 1,200 to 1,900 miles of Long Valley, know that this was the grand finale.

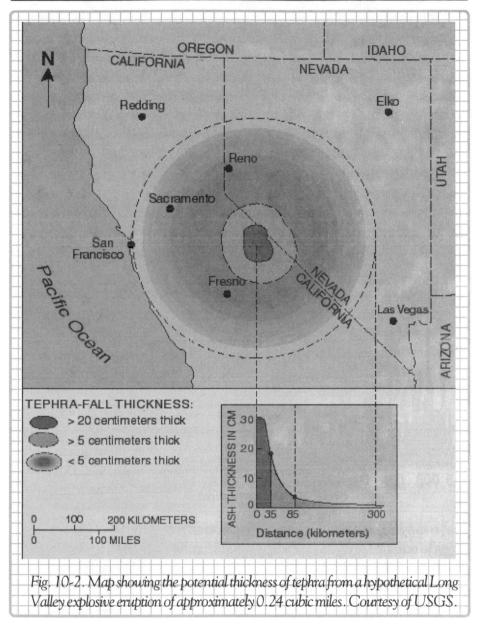

Fig. 10-2. Map showing the potential thickness of tephra from a hypothetical Long Valley explosive eruption of approximately 0.24 cubic miles. Courtesy of USGS.

The sound was horrendous, and could be heard as far as thousands of miles from the caldera in Phoenix, Denver, Salt Lake City, and beyond. The noise level up to distances of 180 to 250 miles was literally deafening—rupturing ear drums of people out in the open who were still in places like Los Angeles, San Diego, San Jose, San Francisco, Reno, and Las Vegas.

Of course, there were no observers around to witness the collapse, and the approximate size of the new caldera eruption, in terms of total volume of erupted material, would not be known for at least a year. It would eventually be estimated that the eruption sequence, which would be referred to as the Long Valley supereruption of 2015, ejected a bulk volume of 120 cubic miles of material, similar to the eruption 760,000 years ago. This total volume was eventually based on estimates of the three components of a large caldera eruption: intracaldera fill, pyroclastic outflow, and tephra fall-out.

The short-term regional and continent-wide effects were pretty much determined by the volume of ashfall, which was conservatively estimated to be 30 cubic miles. A study carried out in 1982 reported in a USGS Open-File Report, computed the potential thickness of tephra as a function of distance from a hypothetical Long Valley eruption of approximately 0.24 cubic miles of magma. The results of this study are shown in Figure 10-2.

The model for the ash distribution in Figure 10-2 predicts that downwind deposits of ash could reach thicknesses of at least 8 inches at a distance of 21 miles and about 0.4 inches at a distance of 186 miles. The authors of the study note that the potential ash thicknesses are based on ash deposits from volcanoes at locations other than Long Valley that involved eruption volumes of as much as 0.24 cubic miles.

An estimate of the thickness of ashfall at Las Vegas, a distance of roughly 230 miles, from a 0.24 cubic mile eruption at Long Valley is about 0.2 inches. Assuming linear scaling and a volume of 30 cubic miles, the estimated thickness from the 2015 Long Valley eruption at Las Vegas was about 25 inches or 2 feet. To appreciate this thickness consider the preceeding figure, which shows a relatively thin 3.5 inch blanket of ashfall at the Clark Air Force Base from the June 15, 1991, eruption of Mount Pinatubo in the Philippines. The distance from Clark Air Force Base to the volcano was only 15 miles. It must be noted that during the eruption, a massive typhoon made landfall and passed about 46 miles northeast of the volcano. It's obvious from the figure that there was considerable blowing and drifting of the ash. Yet, what's amazing about this figure is that the average ashfall deposit at Clark Air Force Base was only 3.5 inches, while an estimated total of between 2 and 2 1/2 feet of ash fell on Las Vegas during the 11-day sequence of eruptions at the Long Valley caldera. Not only did the perpetual lights go out at night along the Vegas strip, but with a little blowing and drifting of ash, the high rollers weren't able to find their limos. In fact, a few of the casinos collapsed under the weight of ash on the rooftops, as well as some of the many surrounding hotels, motels, and out of the way places Vegas is known for. A year or so later, Pompeiified (preserved in ash similar to the residents of Pompeii in A.D. 79 following the eruption of Vesuvius) slot players were found sitting at their Double Diamond machines staring vacantly at the always occurring mismatch of symbols, and clutching their plastic cups half full of nickels. Obviously, what happens in Vegas would stay in Vegas, including the money, for a while, at least.

Fig. 10-3. An ashfall deposit of approximately 3 1/2 inches at the former Clark Air Force Base on June 16, 1991, from the Mount Pinatubo eruption in the Philippines. Clark was located about 15 miles east of the volcano.

The air in the states closest to the eruption of Long Valley was full of ash, rendering it unbreathable. Those who hadn't already evacuated at the first sign of unrest were forced to make the terrible choice between staying indoors and running out of food and drinkable water, or venturing outside and succumbing to a slow death by choking on the tiny, but deadly ash falling like rain upon a grey, sunless landscape. It would be like drowning in a liquid form of concrete—the deadly Marie's Disease suffered by mammals more than 10 million years ago in and near a watering hole in the modern-day state of Nebraska. Microscopic shards of ash, akin to tiny bits of glass, cause the lungs to be filled with blood, causing a miserable form of choking to death.

All airline routes in and out of the western United States are suspended, and any Air Force jets stationed out west that could be evacuated before the initial eruption blackened the skies are sent to points east, to sit tight until the initial chaos was over. Following the caldera roof collapse, the ash cloud continues to spread, covering up to 80 percent of the nation in a bed of ash, leaving approximately 20 percent totally uninhabitable to human or animal life.

The president continues to broadcast constant news conferences, trying to keep the surviving populace from breaking out into a panicked state of anarchy as confusion mounted over how best to help them. FEMA consults hourly with the Pentagon and top Homeland Security officials about military assistance, as food-related riots break out all over the nation, even in states relatively stocked with supplies.

Within days and weeks of the supereruption, the suspension of air routes, the inability to bring cargo in and out of the most deeply affected areas, and the virtual decimation of the Grain Belt, the area of our nation responsible for the vast majority of our grain food sources, all contribute to a growing sense of desperation and panic among survivors anxious to find food. It only takes 0.04 inches of ash to close airports, and the wide swath of blanketed ash would literally shut down every major and minor airport for thousands of miles across the country.

Because even a small amount of ash can clog an engine, road transportation is heavily curtailed, and trucks and machines normally engaged in the moving of supplies from one state to another find themselves immobilized. The transport of cargo across borders ceases, both for logistical and safety reasons. Electrical equipment shorts out, and wide areas experience power outages and rolling blackouts, rendering communication via computers and phones obsolete. Ham radio operators once again become the most likely source of getting information in and out of the worst hit areas, and shortwave radios became hotter than the most cutting edge mini-handheld electronic devices that have just hit the market.

In the states closest to the eruption event, more and more food riots erupt as hungry survivors clamor for what little supplies remained on shelves of stores, and the quest to find an even greater rarity, noncontaminated drinkable water, leads to violence and growing anarchy. People begin roaming the streets, armed to the teeth, with a survivalist mentality driving them to violence over something as simple of a jug of water or a package of cookies.

Unfortunately, there is little government stability to respond. The president's previous Major Disaster Declaration, which would provide for long-term federal recovery assistance to those in need, just doesn't resonate with the millions of people whose sole thought was to find food, shelter, and water to get through another day. It's hard to think in the long term when you may not live to see tomorrow. Disaster aid meant nothing to those watching their children die of starvation, the millions left homeless and without direction, forced to find shelter in overcrowded and understaffed shelters. Fewer and fewer people turn to the government for assistance, and for hope, as the conditions around them worsen, even as the eruption sequence itself dies down.

Forced to declare Martial Law, and suspend individual Constitutional rights, the president puts the National Guard and military police system into action. With the world around them falling to pieces like the ash falling from the darkened skies, most people move into a complete survival mode, with the reptilian brain taking over all semblance of common sense, or common good. And there simply is not enough military police in the nation to stop them from finding a way, any way, to survive and protect themselves and their families.

In the days and weeks following the Long Valley supereruption, it is every man, woman, and child for themselves.

The intermediate and long-term continental and global effects are determined by a combination of the ash and the volcanic gases (aerosols) lofted into the stratosphere. Within two weeks, the ash and aerosol cloud circled the globe, and is beginning to spread toward the southern hemisphere. The cloud causes significant dimming of the sun, at times reducing the solar radiation by 90 percent, for several weeks, which adversely affects photosynthesis of plants. Even after the ash particles have settled out of the atmosphere, the aerosols remain suspended in the upper atmosphere for several years and, starting with the summer of 2015, reducing global temperatures by 5 to 11°F. Thus, the volcanic winter last seen after the Toba supereruption would be with us for the next three to five years, creating havoc with our ecological and environmental well-being.

As the global skies dim, and demand for food and water increase, those left behind would be forced to contend with the rotting corpses of the dead, both human and animal. Millions of wild and domestic animals such as cattle and foul raised for food litter the landscape, creating hotbeds of potential disease. Water contaminated by ash tempts those dying of thirst, even as it contributes to the spread of diseases and illnesses that challenge what little strength remains in the surviving population.

Cargo ships are unable to get into the major ports along the west coast, shutting off food, water, and medical supplies to the immediate vicinities most in need. Those survivors in states further from Long Valley see slow federal assistance response, as the Pentagon, FEMA, and even the Red Cross struggles to first locate supplies, and then find a way to get them into the zones of highest need, most likely with targeted air drops, if skies permit. It becomes a game of survival triage, with the few able responders tending to thousands, if not millions of injured and dying.

With hundreds of thousands of survivors trying to leave the United States, we might face a reversal of border closings, with both Canada and Mexico forced to turn back the human tide. These nations do assist the U.S. with food and supplies, but are dealing with the global affects of the supervolcano themselves, with drastically reduced crops and food supplies curtailed by the suspension of trade routes. The U.S. dollar, already crippled by a slow economy and increased influence of the European and Asian markets over the last 20 years, most likely dies along with thousands of victims, buried under the collapsing economy just as the thick ash would bury innocent victims under the weight of collapsed roofs and caved in buildings. Wall Street ceases to exist, with a crash never before witnessed, even by those related to survivors of the Great Depression of the preceding century. Nations that relied on the U.S. for exported goods are forced to turn elsewhere, and no longer is the U.S. the leader of the world, as other nations less affected step into power, touting their own strong monetary system, food, and water resources, and abundance of oil and gas. Where the U.S. once stood in military power, new nations rise to take our place, taking advantage of the crippled eagle that once flew free and mighty.

Many futurists have speculated that a major disaster such as a supervolcano might herald the rise of Third World countries to geopolitical power. Even though all nations of the world would be affected by the Long Valley supereruption, the catastrophe would clearly tip the balance of power far away from the West. Latin American nations coalesce into a powerful new force, with rich natural gas and oil supplies not yet tapped out during peak oil years, and with military force mostly purchased from their neighbors to the north. The European Union increases its own influence in the global body politic, as do the nations of the East such as China, India, and Pakistan, and new alliances between Russia and Middle Eastern nations such as Iran and Syria create a new superpower. The nations least affected by the drifting ash become the New World Order, with a complete reshuffling of military might and economic clout.

It would be years, perhaps decades or more, before the United States would be close to regaining its power and position in the world.

In Europe, people witness the phenomenon of red skies, due to the gasses and sulfur in the air. Even areas with only a minimal layer of ash become danger zones when rainstorms break out, creating torrential flows of water and ash, similar to a thick and fast-moving mud, that creates flooding and devastation in its path.

In nations across the globe, various aftereffects manifest, but most of the remaining populace of the planet would be glued to news about the United States, as it plunges into anarchy, struggling to survive the immediate and long-term consequences of a caldera that felt the sudden need to express itself.

It will be at least three months before transatlantic routes once again open to travel to and from the United States. It would be years before a new structure of government takes effect, or before the economy sees any sign of rebirth and regrowth. It might even be a decade before the United States is once again a healthy, strong force in the world. Then again, the results of the supereruption might be so devastating, the people of the United States might never recover, at least not without a drastic change in the way they live and interact with their environment.

People addicted to the magic and ease of high technology are reduced to the sheer, primal act of survival, and suddenly forced to live off the land, and their own wits. Rural people and indigenous peoples thrive, even as their urbanized "evolved" counterparts flounder and die, unable to grow their own food fast enough, or decontaminate their own water quick enough, or avoid the dangers of others, sometimes their own neighbors, who are just as desperate to survive and would kill to do so. Ironically, it will be the very technology we all craved as part of the Information Age that might do us in, with those living close to the earth destined to become the more highly evolved. Our need for speed, for faster and easier and cheaper, becomes our downfall, the future belonging to those we once foolishly labeled "primitive," "backward," and "unevolved."

Even as the survivors, in the months following the supereruption, begin to climb out of the ashes, so to speak, the changes in global climate deal even more blows to their will to survive. The cooling brought about by the ash in the atmosphere leads to a new mini ice age, wiping out crops and food sources in the northern hemisphere, and crippling production in other areas that are soon forced to feed a growing number of humans, with far less output.

Agricultural regions are devastated by the change of temperatures, sometimes as drastic as a 9 to 18°F drop, and the extreme crash of food production sources no doubt creates further political ramifications for a world already on the brink. Add to this the already increased demand, from sheer human overpopulation in the new millennium, for drinkable water, and the makings of a world war would be like crisp, dry kindling being tossed upon an exposed flame. The very fabric of civilization begins to tear under the stress and strain of severely curtailed production of the most basic needs of survival. The deterioration of the global climate lasts for years, bringing with it more devastation and death on a widespread scale as some of the already poorest nations struggle with mass population loss due to decreased food imports, especially grain, and watch as more powerful nations take control of the few major water sources for their own overwhelming needs.

Ironically, the supereruption would put a stop to the rampant global warming that was threatening the existence of humanity, and the resultant cooling forces nations to reconsider their entire strategies for dealing with the rising temperatures and sea levels. Now, they have to think in terms of colder climates, icier regions, and far less sunlight reaching the Earth than any had ever witnessed in their lifetimes.

Volcanic winter, akin to nuclear winter, is even more deadly than the initial event that brought it into existence. More people perish in the coming two to six years duration of the volcanic winter period from lack of food, water, shelter, disease, warfare, and struggle over dwindling resources, and from emotional stress and the inability to cope and work in groups for the benefit of all….More perish from this post-eruption period than during the initial eruption itself.

There is still one more, even deadlier, effect that awaits the survivors of the volcanic winter. This effect, addressed in Chapters 5 and 9, is the depletion of the ozone layer and the bombardment of life just starting to emerge from the trials and tribulations of the previous volcanic winter by deadly UV-B radiation. Ironically, the severity of this problem increases, as the atmosphere was cleansed of ash and aerosols, and people begin looking forward to the end of their ordeal. The increase in atmospheric transparency allows the ultraviolet radiation unfettered access to the Earth's surface. Excessive exposure to UV-B radiation can result in a wide range of damaging health effects to humans and animals. These effects include skin cancer, severe eye damage (cornea, lens, and retina), and impairment of the immune system. Once this becomes widely known, people are forced to either cover themselves as much as possible during the daylight hours, given the unavailability of sunscreens, or become nocturnal.

In the years following the end of the volcanic winter, and the gradual return of the ozone layer to concentration levels prevailing prior to the supereruption, those who survived begin to see the birth of a new kind of civilization, one built upon a greater respect for nature and natural disasters, and upon the knowledge that technology doesn't always serve to ensure survival. The survivors of Long Valley would go on, rebuilding communities, creating new jobs, and a sense of hope that always comes out of a disaster, no matter how big or small.

We survived Toba, and that supereruption narrowed us down to just a few thousand hearty, and possibly lucky, souls. A supereruption in 2015, or even next week, would end the lives of millions, but the species would go on. It might take decades before a workable economy is back in place, and our military might once again assures our position in the world. It might take decades before we can ever think about pursuing new technologies that make life easier, so focused we will be on just living day to day. And it might take decades before people will return to some sense of calm and normalcy, although what will be normal 10 years after a supereruption will be quite different from what we think of as normal today.

It might be decades before we can go to sleep at night without fear of never waking up the following morning, or go to our jobs with the sense of purpose and meaning we had before it all changed, before the caldera blew its top and imploded in on itself—a simple act of nature....

But we, the authors, believe that despite the dark days that would follow the eruption of Long Valley, or Yellowstone, or even Toba again, we would continue to move toward the dawn of a new day.

Because the human spirit is stronger than any supervolcano on Earth.

CHAPTER 11

Are We Prepared?

Can one be prepared for something as catastrophic as a supervolcano? Perhaps if the next big supereruption occurs in our own backyard, there is little we can do to increase our chances of survival. But if that event should occur across the globe, can we, indeed, survive the environmental, emotional, and economical collapse that is sure to come?

The devastation caused by Hurricane Katrina proved that even the most powerful government in the world could not contain the damage of a Category 3 monster storm making landfall. The chaotic response, and in many cases, utter lack of response, is still being felt to this day by residents who have, despite the media moving on to bigger, brighter things, not recovered their lives, or their property. Parts of the Gulf still are in shambles. Lives are still in disarray.

Prior to Katrina, the September 11th terrorist attacks brought to the forefront all that was, and still is, lacking in homeland security. Threats of a bird flu pandemic, or another terrorist attack, or more massive hurricanes and superstorms fueled by both natural and human-caused global warming have certainly pushed disaster preparedness to the forefront.

But a supervolcano far surpasses any superstorm or terrorist attack, or even pandemic, in the sheer volume of death and destruction it can cause.

FEMA, the United States Department of Homeland Security's Federal Emergency Management Agency, initiates a Disaster Process, followed by Disaster Aid Programs, when a catastrophe occurs. This Process is as follows:

✳ **First Response** occurs when a disaster strikes. Local government emergency services, and state and volunteer agencies kick into gear, followed by federal resources mobilized via request from the governor. FEMA then enters the picture for search and rescue, food, water, and shelter provisions, and electrical power needs.

✳ **Long-term recovery** is the next phase, and the most costly, taking into account damages to public facilities and infrastructure. At the governor's request, the area can be considered for major disaster funds and recovery efforts. This happens often after major quakes, flooding, and hurricanes/tornadoes.

✳ **A major disaster** would be big enough to warrant presidential intervention, in the form of additional federal aid. This is what occurred after Hurricane Katrina and, of course, the September 11th terrorist attacks. A presidential Major Disaster Declaration would initiate long-term federally funded recovery programs to help disaster victims, businesses, and public facilities.

✳ Then, there is an **emergency declaration**, which is more limited in scope, and often involves less long-term aid. An ED is often utilized to help stop a major catastrophe from occurring, as in dam or levee repair, which certainly the Gulf region could have used before Katrina struck.

With a supervolcano, surely a Major Disaster Declaration would take place, along with Emergency Declarations for areas not immediately affected by the initial eruption and/or pyroclastic flow. First, local governments would respond with help from volunteer agencies. State assistance would then be called upon, if needed, and possibly the National Guard and other state agencies would be brought into the picture.

Together, local, state, federal, and volunteer organizations would perform a detailed Damage Assessment and discern needs that might require a federal response. The governor might then declare the Major Disaster Declaration, requesting help from the president, who would then send out FEMA to evaluate the situation. If the president approves the FEMA evaluation, he/she would then declare federal emergency assistance.

On an individual level, aid could come for survivors in the form of short-term housing or funds, such as disaster grants and low-interest loans, unemployment assistance for those left jobless, property damage inspection, and a host of programs to secure help with legal aid, taxes, and government benefits.

Public Assistance would help rebuild community infrastructure, and work on a state and local level. In most cases, PA programs pay 75 percent of the project's costs, and can include everything from removing debris to providing extra security measures.

Hazard Mitigation allows disaster survivors and public entities to take steps to avoid loss of life and property in future disasters, and includes earthquake reinforcement and retrofitting; dam and levee enforcement; and enforcement of local and state codes.

But the process only works if you survive those first 72 crucial hours that emergency services urge us to be prepared for, and, as we've seen in previous chapters, those living within the reach of a supervolcanic eruption will most likely not survive, unless they happen to be underground or living in a protected dome with its own source of fresh food, water, and air.

For those on the perimeters, and those beyond the initial grasp of the immediate damage, survival will depend on many things, including proactivity before the disaster strikes, and the ability to get as far away from the event area as possible in as quick a time period as possible. Even then, it may not be enough.

In the previous chapter, we speculated on the chain of events that might occur should Long Valley's mighty caldera decide to release their magma pressure buildup. Those living near less destructive volcanoes, such as Mt. St. Helens, can indeed take steps to assure a greater ability to not just survive the eruption, but come out as unscathed as possible. After the eruption of a "normal" volcano, even a big one such as Krakatoa or Pinatubo, life will, and does go on.

The USGS provides great resources for getting through a volcanic eruption, with tips for what to do before, during, and after the event. These tips make sense for those who would not be immediately affected by a supervolcano, but only in the short term, because again, as we have learned, it is the aftereffects such as climate change, and even a mini ice age, that would render much of these useless.

But the scale is always balanced in favor of preparation.

WHAT TO DO BEFORE, DURING, AND AFTER A VOLCANIC ERUPTION

BEFORE

1. Learn about your community warning systems and emergency plans.

2. Be prepared for additional disasters that can be triggered by volcanoes:
   * Earthquakes.
   * Flash Floods.
   * Landslides, rockslides, and mudflows.
   * Thunderstorms.
   * Tsunamis.

3. Make evacuation plans. Get to high ground and away from eruption. Have a back-up route in mind.

4. Develop an Emergency Communication Plan with family members if separated during eruption. Ask an out-of-state relative to be a contact person that you can all call into. After a disaster, it is usually easier to call long-distance than locally.

5. Have disaster supplies on hand.
   * Flashlight/batteries.
   * Portable radio/batteries.
   * First aid kit/manual.

* Emergency food and water.
* Manual can opener/knife.
* Medicines/prescriptions/eye glasses/documents/insuance papers.
* Cash and credit cards. (Cash is best!)
* Extra clothing and shoes.
* Goggles and breathing masks for household members.
* Pet needs.

6. Contact your local emergency management office or Red Cross chapter for more information on volcanoes.
7. Be prepared to evacuate. Know what you will take ahead of time!

DURING

1. Follow evacuation orders issued by authorities. Don't panic—it does not move things along, but bogs things down.
2. Avoid areas downwind of the volcano.
3. If caught indoors, close windows, doors, and dampers. Put machinery inside barn/garage. Bring livestock into closed shelter.
4. If trapped outdoors, seek shelter. Avoid low-lying areas where poisonous gases can collect and flash flooding is a threat. Beware of mudflows, falling rock, and debris.
5. Protect yourself with heavy clothing, goggles, dust mask, or damp cloth. Keep car and truck engines OFF.
6. Stay out of main event area! Being a looky-loo will get you killed.

AFTER

1. Listen to battery-operated radio or TV for latest emergency information.
2. Stay away from ash fall.
3. Cover nose and mouth if outside, wear goggles to protect eyes, keep skin covered to avoid burns and irritation.
4. Stay indoors until help arrives if you have respiratory problems.

5. Avoid driving in ash fall—it can clog engines and stall vehicles.
6. Clear roofs of ash fall to avoid collapse.
7. Help neighbors, especially infants, the elderly, and the disabled.

Adapted from FEMA for USGS.

Both FEMA and the USGS offer additional information about dealing with the main problem of a supervolcano, for those not living in the immediate eruption zone: ash fall. Because ash fall can block the sun, most likely people will be operating in the dark, with greatly reduced visibility. Lightning often accompanies ash fall, creating even greater dangers for those stuck outdoors with little shelter.

Ash damages machinery and disables transmission lines, causing electrical short circuits. Chances are, following a supervolcano, there will be widespread, and increasing, reports of power outages. Water supplies and drains will be clogged and contaminated. Living three days without power and fresh water is difficult enough. Imagine going for weeks without it, even months. We can ask the Katrina survivors about it. Many of them are still living in such horrid conditions, more than a year after the terrible storm struck the Gulf region.

Because ash does not melt, such as snowfall does, it accumulates on rooftops and carports. A 1-inch layer of ash weighs 5 to 10 pounds per square foot when dry, but between 10 to 15 pounds per square foot when wet. Roof and building collapses will be numerous, and a clear and present danger to those staying inside to avoid the dangers outdoors. In a way, you're damned if you stay inside, and damned if you go outdoors.

Ash, when wet, is incredibly slippery, and will be chaotic for evacuation procedures involving any kind of vehicle. Roadways may be blocked with debris and rocks, mudflows, or perhaps flash flooding in low-lying areas. But if those threats don't materialize, it is still the ash one must contend with.

If mechanical failure is a probability, keeping a carload of extra clothing, blankets, and food and water might increase chances of survival, and at least make things a bit more comfortable, until help can arrive. The USGS site also suggests keeping an extra air filter or two in the car if you live near a volcano or caldera, as ash will soon clog the existing filter and render the car inoperable.

After a supereruption, one of the main long-term problems becomes finding fresh food and water sources, and those fortunate enough to have a garden or personal water supply will have to deal with ash contamination. Washing food from the garden is fine, if you have clean water to wash it with. FEMA suggests that, if no clean water is available, you try letting ashy water sit until the ash settles, then use the top water. Those who own pets and livestock will also have to provide food and water for them as well. It could, as the previous chapter indicated, be weeks or even months before infrastructure is rebuilt and placed back into some semblance of service, allowing for the transport of fresh food and water supplies into a devastated area.

Those living in valleys must then contend with mudflows and flash-flooding. Volcanic mudflows (called Lahars), can carry large boulders and have the thick consistency of wet concrete. Mudflows move quickly—faster than a person could run—but a car can outpace a mudflow, if travel by car is an option. As distance increases from the volcano, so too does the threat of mudflows. It really is all about getting as far from the event as possible, and as quickly as possible.

Floods remain the most common and widespread natural disaster plaguing the United States, and after a volcanic eruption, they are a grave threat. Just 2 feet of fast-moving water can sweep a car away. Flood warnings would be immediately issued following an eruption, but for those who may not have the means to hear or see the warnings, the idea is a simple one: get away from any water source. Find out beforehand if there are flood-prone areas and avoid them at all costs. Think about the many news stories showing people trapped in their cars as they attempted to drive through rising waters. It doesn't work.

Obviously, the more preparation one does before a disaster strikes, the better, but people rarely, unless they live in an area frequented by earthquakes or hurricanes and tornadoes, think in such proactive terms. Even those who do live in high danger zones get lazy and complacent. Should a supervolcano threaten to erupt, hopefully we will have some warning, and there will be time to do some of the things needed to ensure survival, or at least level the odds.

Surprisingly, the most prepared areas of the nation may be the rural areas, according to a January 2007 article in *EurekAlert* titled "Rural America More Prepared for Disaster…Also More Vulnerable." Rural sociologist and professor Courtney Flint and one of her students, Joanne Rinaldi, interviewed 20 coordinators of Community Emergency Response Teams (CERTs) and found that the rural CERTs live closer to the land, which makes them more prepared in many ways, including access to generators, kerosene heaters, snow plows, and other necessary equipment suburbanites and city dwellers don't have. Rural CERTs tend to plan ahead, and plan as if they themselves will be the first responders to a disaster. They tend to have a more "we can't wait for help so we have to do it on our own" attitude, unlike their more urban counterparts, who have access to first response police and fire teams that may not make it into rural areas.

But the isolation of rural areas adds to their vulnerabilities, which is often offset by the ability of rural communities to become more self-reliant.

Surviving a supervolcano will come down not to the survival of the fittest, but of the smartest and most prepared; not to mention, the most psychologically resilient. A major factor post-disaster is recovering emotionally from the trauma of a catastrophe. Disaster-related stress begins with the shock of first witnessing an event, especially for those directly effected. On the morning of September 11th, people across the United States, indeed the world, went into shock, shutting down emotionally, and even physically, as the events of that day unfolded. Those living in New York City and working in or near the Pentagon were even more deeply affected.

Those effects may have waned through time, but still are prevalent in higher rates of generalized anxiety, depression, fear, and feelings of insecurity. *Post-traumatic stress disorder*, discussed in detail in Chapter 8, continues to manifest, and not just in the soldiers returning back from Iraq years later, but in the citizens who still remember that fateful morning.

As we saw in Chapter 8, there are signs of disaster-related stress that people must be aware of, especially in regards to children. These signs include:

✳ Difficulty communicating thoughts.

✳ Difficulty sleeping.

✳ Irritability.

✳ Depression, even suicidal thoughts.

✳ Increased use of drugs, alcohol, or prescription meds.

✳ Poor work performance.

✳ Lack of concentration.

✳ Headaches/stomach and digestive problems.

✳ Reluctance to leave home.

✳ Fearful thinking.

✳ Paranoia.

✳ Mood swings.

Just as in any other traumatic event, people can find help after a disaster through the local, state, and government agencies, especially the Red Cross, as well as faith-based agencies, churches, and religious places of worship, and counselors and therapists. This is especially critical when dealing with children, who may not be able to properly process the events and lack coping skills.

Asking for and accepting help is the first step toward emotional recovery, and is difficult when one is also immersed in trying to find food, water, and shelter. Family, friends, and communities offer the best opportunities for reaching out, but if they are not available, or cannot be contacted after a catastrophe, it becomes a true test of personal strength and resilience.

The National Center for Post Traumatic Stress Disorders suggests witnesses and survivors adopt the **Protect, Direct, Connect, Select** strategy:

✳ **Protect:** Find a safe haven that offers food, shelter, water, sanitation, and the chance to get quiet and rest.

✳ **Direct:** Begin working on immediate personal and family priorities to preserve a sense of hope, direction, and purpose.

✳ **Connect:** Maintain and reestablish communication with family, friends, neighbors, and people you can talk to about your experiences, especially counselors or faith-based spiritual guides.

✳ **Select:** Identify key resources such as FEMA, the Red Cross, Salvation Army, and other local and state organizations that can provide housing, clean up, and basic emergency needs.

FEMA suggests that one of the best ways to cope after a catastrophe is really one of the best ways to avoid some of the trauma in the first place, and that is to stock up and prepare for the next time. Being proactive, even after the fact, has its merits, and helps to engage the mind, redirect anger and fear into positive action, and help everyone band together to make sense of the brutal assault on the Earth, and on their lives.

No matter where the big one strikes, there will be survivors. Human ingenuity is already attempting to find ways to make sure of it. In 2001, the Grace A. Bersted Foundation provided a grant to start a program called SEMP, the Suburban Emergency Management Project. SEMP focuses on the local perspective for disaster preparedness, response, recovery, and mitigation emphasizing, through education, community-based needs. As more and more local level organizations take responsibility for helping their own communities get through major disasters, it will leave the state and federal agencies a bit freer to help in massive and catastrophic events, such as a supervolcano.

In a February 9, 2007, article for Agence France Presse, reporter Penny MacRae unveiled the "Doomsday Vault," a plan to help resist global warming effects. This Arctic repository will be carved into the permafrost of a mountain in the Svalbard archipelago, a remote area near the North Pole. Designed to sustain the effects of climate change, the vault will preserve and grow approximately 3 million batches of seeds from every imaginable crop on Earth.

Also called the "Noah's Ark of food," the project will be situated about 426 feet above sea level, and is expected to be safe and secure even if the Arctic ices melt completely. Construction is due to be completed in late winter of 2008.

Perhaps we should be thinking in terms of more "Doomsday Vaults," in the ice, underground, below mountains, and any other place we can build them, to store seeds, foods, clean water, even clothing. Because when the big one hits, we will need more than just one to get us through the global climate change from the years of volcanic winter. But the concept is a start, and should scientists begin to raise the volume of concern over a potential supereruption, the construction of these vaults, and perhaps even domed shelters to keep people safe from the ash, climate, and cold, will become a part of a world-wide disaster preparedness plan.

NASA is preparing to test inflatable lunar shelters to be set up as outposts on the Moon. These inflatable structures are made of multilayer fabric and are approximately 12 feet in diameter and 18 feet tall. The shelter is connected to an inflatable airlock, connected by an airtight door. Eventually, NASA hopes to develop the shelters to contain sleeping quarters, walls, and floors. Perhaps these inflatable shelters could be produced en masse to provide survivors of a supervolcano with a roof over their heads. The question is, would anyone buy one? It's hard enough to get Californians to put together an earthquake kit despite numerous temblors serving as a constant reminder of the Earth's instability.

In the July-2006 issue of *Reader's Digest*, an article by Alice Lipowicz titled "The Next Disaster: Are We Ready?" examined the preparedness of the 10 most high-risk U.S. cities. The expectation was that after the September 11th attacks and Hurricane Katrina, our nations' cities would be ready for the next big calamity—possibly another attack, a bird flu pandemic, or a massive West Coast earthquake. The report found that indeed, every major urban center had received federal funding, the majority of it from the Department of Homeland Security, to make cities more secure.

But this same level of security did not apply to preparedness, an area the federal government is working hard to catch up on. The report looked at several key issues:

1. Emergency Readiness: the level of first responders per 100,000 residents. First responders include police, fire, and EMTs.
2. Federal Search-and-Rescue Teams within 50 miles of each urban center.
3. "Green Status" from the Centers for Disease Control and Prevention. Green status means the area will be able to quickly mobilize in a pandemic or bio terror attack, and will have a stockpile of vaccines and medical equipment to meet the need.

4. A city Website that fully explains evacuation procedures, including details for people with special needs.

5. Crisis Communications—can first responders talk to one another?

6. E911 or Enhanced 911 technology.

7. Medical Response of at least 500 hospital beds per 100,000 residents.

8. Local response teams trained to work together.

9. Nearby labs specializing in biological and chemical threats.

Of the 10 urban centers studied, many fell within the accepted criteria range, with some cities rating above standard on some issues, and below on others. In general, the highest score a city could get was 100 percent, this by totaling the points earned in each category and then converted to a percentage formula.

The highest rated urban centers are Miami, New York, and Washington, D.C. In the middle, with headway in some areas and lack of preparedness in others, are Boston, Chicago, Houston, and Los Angeles. Scoring at the bottom are Philadephia, Las Vegas, and, the most unprepared city of all in the study, Detroit.

Sadly enough, even years after September 11th and Hurricane Katrina, most of our urban centers are not fully prepared for a small- to large-scale scenario. Imagine, then, how prepared they will be for a supervolcano. In a December 2006 *USA Today* report by Mimi Hall, the consensus was that most Americans are simply unprepared, no matter where they come from. This article focused on the state of preparedness in Florida, and despite consistent and constant messages about preparedness, the reporter commented that most of the residents simply do not listen. The non-profit Council for Excellence in Government developed a Public Readiness Index, rating preparedness on a scale from one to 10, based upon answers to 10 questions.

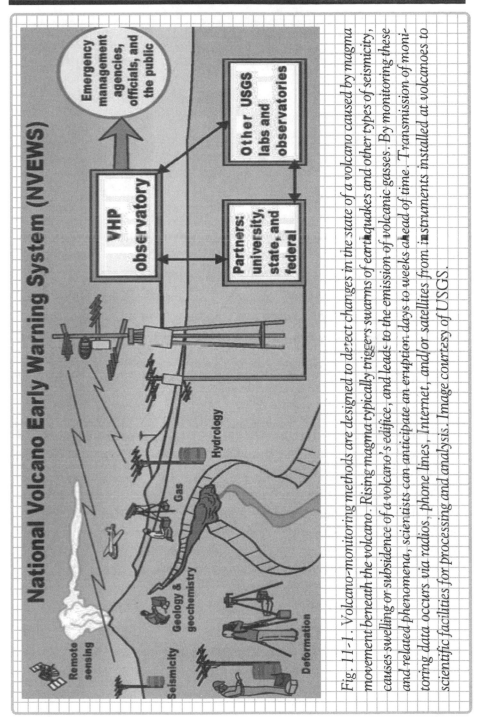

Fig. 11-1. Volcano-monitoring methods are designed to detect changes in the state of a volcano caused by magma movement beneath the volcano. Rising magma typically triggers swarms of earthquakes and other types of seismicity, causes swelling or subsidence of a volcano's edifice, and leads to the emission of volcanic gasses. By monitoring these and related phenomena, scientists can anticipate an eruption days to weeks ahead of time. Transmission of monitoring data occurs via radios, phone lines, Internet, and/or satellites from instruments installed at volcanoes to scientific facilities for processing and analysis. Image courtesy of USGS.

Sadly again, the results showed that people 65 years of age or older, some of the most vulnerable, are much less likely to be prepared than young adults; people with higher education levels are more prepared than those with less education; parents of children who have emergency plans at schools are more prepared, but most parents don't even know of the school's plans; and Hispanics are less prepared than whites and African-Americans.

These facts are not presented to frighten or terrify, but to activate interest and motivate action. Survival begins at home, and then spreads out to the local community, then to the region, the state, and the nation. Planning for a supervolcano may be a moot point for those living too close to the caldera, whether it be Yellowstone or Long Valley or even Toba, but for those who do survive, it will be due to a combination of preparing for the worst, and being physically and psychologically ready to act, should the worst ever happen.

Obviously, the word has not quite reached the public en masse. But the USGS is working hard on ways to plan for any kind of volcanic eruption, and is developing the National Volcano Early Warning System, NVEWS, to work in conjunction with local, state, and national organizations, as well as universities and other USGS labs.

NVEWS was the focus of a major series of meetings held in 2004 by the USGS, and the hopes were high that NVEWS would be a viable tool in the very near future for helping to alert the public in a time when a major eruption is imminent. In addition, the USGS has many volcano-monitoring observatories at such locales as Long Valley and Yellowstone, where scientists can keep constant watch over any developing activity.

The science of forecasting volcanic activity has seen many major advances over the last two decades. Scientists now understand that there will be signs, that the buildup prior to an eruption will be detectable for weeks, even months, beforehand, giving them ample time to warn the public. By constantly monitoring strong earthquake swarms, rapid ground deformation and other types of activity, scientists may be able to give us, the public, days, even weeks, to prepare.

Supereruptions would, of course, be preceded by even bigger precursory activity, which is why the USGS has set up the Yellowstone Volcano Observatory (YVO) and the Long Valley Volcano Observatory (LVVO) to focus on the activity at two of the most potentially hazardous calderas on the planet. As part of the USGS Volcano Hazards Program, these two observatories compliment existing observatories in Hawaii, Alaska, and the Cascades region of the Pacific Northwest.

Stephen Sparks of the University of Bristol, in a statement for *LiveScience* in March of 2005, warned about the inevitability of a supervolcano. "Although very rare, these events are inevitable, and at some point in the future humans will be faced with dealing with and surviving a super eruption."

The human species has survived near-extinction before. Toba almost wiped us off the map. And though we now lack the closeness to the Earth, the ability to live off the land, and the understanding of survival of our danger-ridden ancestors, we have the benefit of amazing technology, advanced knowledge, and the power of people working together in times of great duress.

We can, and will, survive.

CONCLUSION

Why Toba Matters

The story of Toba is the story of humanity, how we evolved, and where we came from, and it is the story of survival amidst the greatest of odds. An event that happened so long ago may seem trivial to us as we go about our day-to-day lives, with our worries and needs placed firmly before us, with little cause to look so far back into the past, until we remember that each person we come into contact with in the course of our day is, in a sense, a survivor of Toba. Our bodies may be different, but our DNA tells us we are all related.

As the song by Sister Sledge tells us, "We Are Family."

To be fair, we must also mention that volcanoes, even supervolcanoes, do provide people and the planet with benefits. Without volcanic activity, we would not have the Hawaiian Islands. Hawaii is the only state in the United States built entirely of volcanic materials. Without volcanoes, we would not have some of the most fertile soils on Earth, soils rich in volcanic materials that have physically broken down over millions of years to produce land which, when cultivated, provides for lush vegetation, prime agricultural production, and even the best rice-growing land in regions of Indonesia near active volcanoes. The western United States agricultural regions are almost entirely built upon soils wholly or largely volcanic in nature.

Volcanoes give us geothermal energy, in the form of the Earth's natural heat and heated ground water once explosed to tremendous volumes of magma that cooled below the surface. In fact, the crust of our planet itself is, in large part, the product of this cooled magma, and eons of volcanic activity. Recently, Hawaii is learning to tap into that volcanic energy in the form of geothermal power plants, and studies done at the Newberry Caldera in Oregon show the potential for up to 13,000 megawatts of energy. Wells drilled at the Long Valley Caldera, a supervolcano, are tapping into the caldera's hydrothermal system to supply three local geothermal power plants.

Even Indonesia, land of massive earthquakes and mighty volcanoes, where Toba once blackened the landscape with ash, is looking to develop its own vast resources of geothermal energy. The island chain literally sits upon what might be the world's largest geothermal resource base, but government bureaucracy has stood in the way of past development for a source of energy that could easily supply the needs of the region's more than 220 million people. Chevron Corporation is currently the only outside business that is developing geothermal plants in Indonesia, and only because they signed a contract decades ago when outside investors were much more welcome, but hopes are high that a new political atmosphere will lead to legislation allowing more foreign investors to develop new geothermal projects in this volcanic hotbed.

Volcanic materials are used to build roads, and even buildings. Solid, lighter volcanic material called pozzualana is mixed with cement and was actually used to built the Suez Canal, as well as many Mexican temples and villages in Italy and France, which often used nothing but this volcanic cement. Volcanic rocks and hardened magma are the source of precious gems such as the opal, zircon, tourmaline, beryl, and moonstone, as well as some of the world's biggest diamonds found in volcanic chimneys in South Africa and the former Soviet Union.

Various metals, gold, silver, copper, zinc, lead, and mercury can be found in the rich deposits of volcanic bombs and cooled lava that has turned to rock. Mount St. Helens in Washington was heavily mined in the late 1800s and early 1900s for copper, gold, and silver, but has long since been abandoned, tapped out by years of prospecting until the 1980 eruption ended mining interests completely. Most of the world's metallic minerals are found within the roots of extinct volcanoes above subduction zones, concentrated around magma bodies by circulating hot fluids that, under certain temperatures and pressures, form rich mineral veins.

We even use volcanic materials as abrasives and cleaning agents, and as the raw materials needed for chemical and industrial uses. New Zealand volcanoes have been quarried recently for basalt and scoria, used in the construction of airports, buildings, and roads.

And of course, volcanoes provide us with plenty of tourist attractions, from Yellowstone National Park, Mount Rainier's Fifth National Park, and the Mount St. Helens National Volcanic Monument, to the Calistoga mud baths of Napa Valley, California, and the hot baths of Rotorua, New Zealand.

Volcanoes, even supervolcanoes, are part of our lives whether we are aware of the notion or not. They create new lands, and rich soils, but they also destroy, and permanently alter the environment. In fact, some scientists believe that life would not be possible without volcanoes. In the January-2000 issue of *The Journal of Geophysical Research*, Washington University geologists Everett L. Shock and Mikhail Y. Zolotov described a scenario in which volcanic gasses may have cooled down into a specific temperature range, between 350 and 579°F, that would have allowed for the prime chemical and environmental conditions for basic hydrocarbons, essential to life, to form. Their work, supported by the National Science Foundation and NASA, suggested that this same cooling process that gives rise to the production of hydrocarbons also allows for the development of amino acids and organic polymers, leading perhaps to self-replicating RNA molecules, and eventually to various types of cells and organisms.

Perhaps volcanic gasses are the reason why we are here, thanks to the eruptions of the last several billions of years, when the researchers suggest the temperature of magma might have been about 400°F hotter than it is today, and the atmosphere less oxidized—a much more agreeable scenario for organic synthesis to occur. Yet, no matter their benefits to life itself, volcanoes remain neutral, indifferent, and objective. We, and we alone, judge them as good or bad, and only in terms of how they affect us.

That a supervolcano such as Toba could erupt today and once again plunge the human species toward the brink of extinction is unlikely, yet it is a possibility, and we must live with that distant and constant threat. Toba did happen, and it, or something of equal devastation, will happen again, and when it does, it will, once again, create catastrophic change that will put humanity on a new path toward a new evolution. Which branch will survive this time? What mutation will occur in our DNA? How will we look 1,000 years later (or more) when the bottleneck ends, and we have evolved into a new line of human beings, far different than what we are today?

But while the threat of a supervolcano cannot be avoided and, should one erupt, cannot be stopped, we face far more pressing threats to the human species today, many of which can be avoided, can be stopped. As this book is being written, thousands of innocent people are dying of poverty, war, terrorism, famine, and violence across the globe—catastrophes we can avoid. We remember that we are family. How could family continue to allow the slaughter in Darfur, or the destruction of the environment, or the poisoning of our seas, or violence toward children? Or perpetual war?

If Toba has taught us anything, it is that we can survive, and that we are all truly one people at the cellular level, where the DNA we each carry links us to one another, even as our outer differences drive us to do horrible things to each other.

Should the world come to an end by means of a natural disaster of epic proportions, a supervolcanic eruption or a massive meteor striking Earth, we have no choice but to accept our fate. But to dwell on that fate is meaningless in light of the ways we might end the world all on our own, without any help at all from Mother Nature. Nuclear war, environmental collapse, economic ruin...

Reinhold Niebuhr wrote the famous "Serenity Prayer" spoken each day by millions of people in recovery across the globe.

"God grant me the serenity to accept the things I cannot change,

Courage to change the things I can,

And wisdom to know the difference."

The story of Toba urges us to accept what we cannot change, such as the mighty supervolcano that may one day erupt, the Toba of tomorrow. It urges us to face, with courage, the things we can change, the things of the world of which we humans do have control of the things we can avoid.

Most importantly, the story of Toba urges us to find the wisdom to know the difference, and to act accordingly. Because, in a sense, we are all in recovery from the catastrophes that have helped shape this planet and its every inhabitants. From the ashes of those catastrophes we have risen.

May the lessons of the past not be lost upon us.

Glossary

Active volcano: A volcano that is erupting. Also, a volcano that is not presently erupting, but that has erupted within historical time and is considered likely to do so in the future.

Aerosol: A suspension, in the air, of liquid droplets or of solid particles.

Aftershock: An earthquake that follows the largest earthquake (or mainshock), in a sequence and originates in or near the rupture zone of the larger earthquake. Aftershocks can continue for a period of weeks, months, or years following the mainshock.

Albedo: The fraction of incident sunlight that is reflected.

Amplitude: The amplitude of a seismic wave is the amount the ground moves as the wave passes by. By way of example, the amplitude of an ocean wave is 1/2 the distance between the peak and trough of the wave.

Andesite: Volcanic rock (or lava) characteristically medium-dark in color, and containing 54 to 62 percent silica and moderate amounts of iron and magnesium.

Ancestral memory: Also known as genetic memory, or racial memory, a collective emotional or psychological genetic transmission of sophisticated knowledge.

Archipelago: A large group of islands, or an ocean area that contains many islands.

Ash: Fine particles of pulverized rock blown from an explosion vent measuring less than 2 millimeters (<0.08 inch) in diameter, ash may be either solid or molten when first erupted.

Ashfall (Airfall): Volcanic ash that has fallen through the air from an eruption cloud. A deposit so formed is usually well-sorted and layered.

Asthenosphere: A ductile layer in the Earth's upper mantle that flows plastically, and where magma is thought to be generated.

Basalt: Volcanic rock (or lava) that characteristically is dark in color, contains 45 to 54 percent silica, and generally is rich in iron and magnesium.

Block: Angular chunk of solid rock ejected during an eruption.

Bomb: Fragment of molten or semi-molten rock, 2 1/2 inches to many feet in diameter, which is blown out during an eruption.

Buckyballs: Molecules composed entirely of carbon, in the form of a hollow sphere, ellipsoid, or tube.

Caldera: The Spanish word for cauldron, a basin-shaped volcanic depression; by definition, at least a mile in diameter.

Cellular memory: A controversial theory suggesting that emotional memory may be stored in cells just as DNA stores genetic information.

Cinder cone: A volcanic cone built entirely of loose fragmented material (pyroclastics).

Coalescence: The merging of two genetic lineages in backward time.

Collective trauma: A traumatic psychological effect shared by a group of people, including entire societies and populations.

Composite volcano: A steep volcanic cone built by both lava flows and pyroclastic eruptions.

Continental crust: Solid, outer layers of the Earth, including the rocks of the continents.

Continental drift: The theory that horizontal movement of the Earth's surface causes slow, relative movements of the continents toward or away from one another.

Crater: A steep-sided, usually circular depression formed by either explosion or collapse at a volcanic vent.

Dacite: Volcanic rock (or lava) characteristically light in color and contain 62 to 69 percent silica and moderate amounts of sodium and potassium.

Dense Rock Equivalent: Tephra volume corrected for void spaces and the density of the rock-type that makes up the tephra.

Dike: A tabular body of solidified magma that fills a fracture in surrounding rock.

Dome: A steep-sided mass of viscous (doughy) lava extruded from a volcanic vent, often circular in plane view, and spiny, rounded, or flat on top.

Dormant volcano: Literally, "sleeping." The term is used to describe a volcano that is presently inactive, but which may erupt again.

Dry fog: A fog, or haze, consisting of dry particulate matter rather than droplets of moisture.

Earthquake: Shaking of the Earth caused by a sudden movement of rock beneath its surface.

Earthquake swarm: A series of usually minor earthquakes, none of which can be identified as the mainshock, occurring in a limited space and time.

Ejecta: Material that is thrown out by a volcano, including pyroclastic material (tephra) and, from some volcanoes, lava bombs.

Epicenter: The location on the surface of the Earth directly above the focus, or place at depth where an earthquake originates.

Eruption: The process by which solid, liquid, and gaseous materials are ejected into the Earth's atmosphere and onto the Earth's surface by volcanic activity. Eruptions range from the quiet overflow of liquid rock to the tremendously violent expulsion of pyroclastics.

Eruption cloud: The column of gases, ash, and larger rock fragments rising from a crater or other vent. If it is of sufficient volume and velocity, this gaseous column may reach many miles into the stratosphere, where high winds will carry it long distances.

Eruptive vent: The opening through which volcanic material is emitted.

Evolutionary Psychology: A theoretical approach to psychology that attempts to explain useful mental and psychological traits—such as memory, perception, or language—as adaptations, that is, as the functional products of natural selection.

Extinct volcano: A volcano that is not presently erupting and is not likely to do so for a very long time in the future.

Fault: A crack or fracture in the Earth along which movement occurs. Movement along the fault can cause earthquakes or, in the process of mountain-building, can release underlying magma and permit it to rise to the surface.

Fissures: Elongated fractures or cracks on the slopes of a volcano. Fissure eruptions typically produce liquid flows, but pyroclastics may also be ejected.

Focus: The point within the Earth where an earthquake originates.

Foreshock: An earthquake that precedes a larger earthquake by seconds to weeks and that originates in or near the rupture zone of the larger event.

Fossil: Any trace or remains of a plant or animal preserved in the Earth's crust.

Fracture: A general term for any break in a rock including cracks, joints, and faults.

Fumarole: A vent or opening through which steam, hydrogen sulfide, or other gases are issued. The craters of many dormant volcanoes contain active fumaroles.

Genetic drift: One of the basic mechanisms of evolution, it is the statistical effect that results from the influence that chance has on the survival of alleles (variants of a gene).

Genetic variation: The variation in a population or species, including the nuclear, mitochondrial, ribosomal genomes as well as the genomes of other organelles. New genetic variation is caused by genetic mutation, which may take the form of recombination, migration, and/or alterations in the number, shape, size, and internal arrangement of the chromosomes.

Harmonic tremor: A continuous release of seismic energy typically associated with the underground movement of magma. Harmonic tremors often precede or accompany volcanic eruptions.

Horizontal blast: An explosive eruption in which the resultant cloud of hot ash and other material moves laterally rather than upward.

Hot-spot volcanoes: Volcanoes related to a persistent heat source in the mantle.

Intensity: A measure of the effects of an earthquake at a particular location on humans, structures, and/or the land itself. The Modified Mercalli Intensity Scale is commonly used to rank the intensity of an earthquake from I to XII according to the kind and amount of damage produced.

Intensity scale: A scale consisting of a series of responses to an earthquake such as people awakening, movement of furniture, damage to buildings, and finally, total destruction.

Isotope: One of two or more species of a chemical element that have different mass numbers (the sum of the number of protons and neutrons in the nucleus of an atom).

KT boundary: Time approximately 65 million years ago between the Cretaceous and the Tertiary; also separates the age of the reptiles and the age of the mammals.

Lahar: A torrential flow of water-saturated volcanic debris down the slope of a volcano in response to gravity. A type of mudflow. Also known as a "glowing avalanche."

Lapilli: Literally, "little stones;" round to angular rock fragments measuring 1/10 inch to 2 1/2 inches in diameter, which may be ejected in either a solid or molten state.

Lava: Magma that has reached the surface through a volcanic eruption. The term is most commonly applied to streams of liquid rock that flow from a crater or fissure. It also refers to cooled and solidified rock.

Lava Flow: An outpouring of lava onto the land surface from a vent or fissure. Also, a solidified tongue-like or sheet-like body formed by outpouring lava.

Lithosphere: The rigid crust and uppermost mantle of the Earth; thickness on the order of 44 to 50 miles. The lithosphere is stronger than the underlying asthenosphere.

Love wave: A major type of surface wave having horizontal motion that is shear or transverse to the direction of propagation (travel).

Magma: Molten rock beneath the surface of theEarth.

Magma chamber: The subterranean cavity containing the gas-rich liquid magma which feeds a volcano.

Magnitude: A numerical expression of the amount of energy released by an earthquake, determined by measuring earthquake waves on standardized recording instruments (seismographs). The number scale for magnitudes is logarithmic rather than arithmetic; therefore, deflections on a seismograph for a magnitude 5 earthquake, for example, are 10 times greater than those for a magnitude 4 earthquake, 100 times greater than for a magnitude 3 earthquake, and so on.

Mainshock: The largest earthquake in a sequence occurring closely in time and space. The mainshock may be preceded by foreshocks or followed by aftershocks.

Mantle: A zone in the Earth's interior between the crust and the core that is about 1,740 miles thick.

Microearthquakes: Earthquakes with magnitudes of approximately 2.0 or less; not commonly felt by people and generally recorded only on local seismographs.

Microsatellites: Sometimes referred to as a variable number of tandem repeats or VNTRs, microsatellites are short segments of DNA that have a repeated sequence.

Mitochondria: Rod-shaped organelles that can be considered the power generators of the cell, converting oxygen and nutrients into adenosine triphosphate (ATP). ATP is the chemical energy "currency" of the cell that powers the cell's metabolic activities.

Mitochondrial DNA: Mitochondrial DNA (mtDNA) is DNA that is located in mitochondria. This is in contrast to most DNA of eukaryotic organisms, which is found in the nucleus. Mitochondrial DNA is passed down through the female genetic line.

Modified Mercalli Intensity Scale: The intensity scale currently used in the United States, developed by American seismologists Harry Wood and Frank Neumann. It does not have a mathematical basis; instead it is an arbitrary ranking based on observed effects of earthquakes.

Multiregional Continuity Model: Also known as Multiregionalism, which suggests that premodern humans migrated from Africa to become modern humans in other parts of the world. *Homo erectus*, after leaving Africa, dispersed into the rest of the Old World and regional populations slowly developed into humans.

NVEWS: The USGS's proposed National Volcano Early Warning System.

Obsidian: A black or dark-colored volcanic glass, usually composed of rhyolite.

Oceanic crust: The Earth's crust where it underlies oceans.

Out of Africa Theory: The popular and widely accepted theory that suggests modern humans arose in one place—Africa.

P (Primary) waves: Also called **compressional** or **longitudinal waves**, P waves are the fastest, and thus the first arriving, seismic waves produced by an earthquake. They oscillate the ground back and forth along the direction of wave travel, in much the same way as sound waves, which are also compressional waves.

Paleoanthropology: The study of human origins.

Pele: The most popular volcano deity. The Hawaiian Goddess of Fire who lives in Hawaii's Mt. Kilauea.

Phreatic eruption (explosion): An explosive volcanic eruption caused when water and heated volcanic rocks interact to produce a violent expulsion of steam and pulverized rocks.

Plates: Pieces of crust and the brittle, uppermost mantle, approximately 62 miles thick and hundreds or thousands of kilometers wide, that cover the Earth's surface.

Plate tectonics: A widely accepted theory proposed in the mid-1960s that the Earth's crust is broken into fragments, or plates, which move in relation to one another, shifting continents, forming new ocean crust, and stimulating volcanic eruptions. Most earthquakes occur when and where plates move past each other either at the Earth's surface or at depth.

Plinian eruption: A violent explosive eruption of smoke and ash with a column of material extending into the stratosphere.

Population bottleneck: The drastic reduction of a population size resulting from a major natural disaster or catastrophe, widespread disease, or human interference such as over-hunting or fishing.

Pumice: Light-colored, frothy volcanic rock, usually of dacite or rhyolite composition, formed by the expansion of gas in erupting lava. Commonly seen as lumps or fragments of pea-size and larger, but can also occur abundantly as ash-sized particles.

Pyroclastic: Pertaining to fragmented (clastic) rock material formed by a volcanic explosion or ejection from a volcanic vent.

Pyroclastic flow: Lateral flowage of a turbulent mixture of hot gases and unsorted pyroclastic material (volcanic fragments, crystals, ash, pumice, and glass shards) that can move at high speed (50 to 100 miles an hour). The term also can refer to the deposit so formed.

Rayleigh wave: A type of surface wave having a retrograde, elliptical motion at the Earth's surface, similar to the waves produced when a stone is dropped into a pond.

Rhyolite: Volcanic rock (or lava) that characteristically is light in color, is 69-percent silica or more, and is rich in potassium and sodium.

Ridge, Oceanic: A major submarine mountain range.

Rift system: The oceanic ridges formed where tectonic plates are separating and a new crust is being created; also, their on-land counterparts such as the East African Rift.

Ring of Fire: The regions of mountain-building earthquakes and volcanoes that surround the Pacific Ocean.

S (secondary or shear) waves: Seismic waves that oscillate the ground perpendicular to the direction of wave travel. S waves will not travel through liquids lie water, molten rock, or the Earth's liquid outer core.

Seafloor spreading: The mechanism by which new seafloor crust is created at oceanic ridges and slowly spreads away as plates are separating.

Seamount: A mountain that rises from the sea floor, but does not reach the surface of the water.

Seismogram: A graph showing a representation of the motion of the ground versus time.

Seismograph: An instrument that records seismic waves; that is, vibrations of the Earth.

Shield volcano: A gently sloping volcano in the shape of a flattened dome, built almost exclusively of lava flows.

Silica: A chemical combination of silicon and oxygen.

Strain: Deformation of a solid material as a result of stress.

Stratovolcano: A volcano composed of both lava flows and pyroclastic material.

Stress: Force exerted per unit area of a solid material.

Strike-slip fault: A nearly vertical fault with side-slipping, horizontal displacement.

Strombolian eruption: A relatively low-level volcanic eruption with material ejected hundreds of feet high.

Subaerial: Pertaining to events or processes that occur on or near a land surface.

Subduction zone: The zone of convergence of two tectonic plates, one of which usually overrides the other.

Tephra: Materials of all types and sizes that are erupted from a crater or volcanic vent and deposited from the air.

Toba Catastrophe Theory: The theory proposed by Dr. Stanley H. Ambrose of the University of Illinois at Urbana-Champaign, suggesting that the Toba supereruption of 74,000 years ago was the cause of the population bottleneck that explains the lack of genetic variation in modern humans.

Tsunami: A great sea wave produced by a submarine earthquake, volcanic eruption, or large landslide.

Tuff: Rock formed of pyroclastic material.

Vent: The opening at the Earth's surface through which volcanic materials issue forth.

Verneshots: Catastrophic explosions of gas through cratons followed by hypersonic blasts of rocks into suborbital trajectories. Hypothesis put forward to explain mass extinctions. Mechanism reminded one of the founding scientists of Jules Verne's book *From the Earth to the Moon*, which is about a huge gun that shoots objects into space.

Viscosity: A measure of resistance to flow in a liquid (water has low viscosity while honey has a higher viscosity).

Volcanic conduit: The channel through which magma rises from within the Earth's crust.

Volcanic gas: A natural gas consisting of mostly water vapor (steam) and also including carbon dioxide, sulfur dioxide, hydrogen sulfide, and smaller amounts of other gases, dissolved in magma and released during a volcanic eruption.

Vulcan: Roman god of fire and the forge, after whom volcanoes are named.

Y chromosome: The sex chromosome that determines male gender. In human genealogy, the Y chromosomal is passed exclusively from father to son.

Bibliography

ABC News Online, "Brush With Extinction." September 5, 2006.

Alberto, Attilio D. "Cellular Memory and ZangFu Theory." *www.attiliodalberto.com*. 2006.

Alvarez, Luis W. "Mass Extinctions Caused by Large Bolide Impacts." *Physics Today* 40 (1987): 24–33.

Alvarez, Luis, Walter Alvarez, Frank Asaro, and Helen V. Michel. "Extraterrestrial cause for the Cretaceous-Tertiary extinction— Experimental results and theoretical implications." *Science* 208 (1980): 1095–1108.

Ambrose, Stanley H. "Late Pleistocene Human Population Bottlenecks, Volcanic Winter, and Differentiation of Modern Humans." *Journal of Human Evolution* 34 (1998).

Arens, N.C., and I.D. West. "Press/Pulse: A General Theory of Mass Extinction?" Annual Meeting of the Geological Society of America in Philadelphia, Paper No. 230–1, 22–25 October, 2006.

Baines, P.G. and R.S.J. Sparks. *"Dynamics of Giant Ash Clouds from Supervolcanic Eruptions."* Geophysical Research Letters *32* (2005).

Battaglia, M. P. Segall, and C. Roberts. "The Mechanics of Unrest at Long Valley Caldera, California. Constraining the Nature of the Source Using Geodetic and Micro-Gravity Data." *Journal of Volcanology and Geothermal Research* 127 (2003): 219–245.

Bechtel SAIC. "Yucca Mountain Site Description." Rev 02, 2004.

Becker, L., R.J. Poreda, A.G. Hunt, T.E. Bunck, and M. Rampino. "Impact Event at the Permian-Triassic Boundary: Evidence from Extraterrestrial Noble Gases in Fullerenes." *Science* 291 (2001): 1530–1533.

Benton, Michael J. *When Life Nearly Died: The Greatest Mass Extinction of All Time*. London: Thames and Hudson Ltd, 2003.

Bindeman, I.N., J. Eiler, B. Wing, and J. Farquhar. "Rare Isotope Insights Into Supereruptions: Rare Sulfur and Triple Oxygen Isotope Geochemistry of Stratospheric Sulfate Aerosols Absorbed on Volcanic Ash Particles." Eos Trans. AGU, 87(52), V23G-08, 2006.

Bindeman, I., J. Eiler, and A.M. Sarna-Wojcicki. "Oxygen-17 Excesses in Products of Stratospheric Volcanic Eruptions and Depletion of the Ozone Layer." Abstract of paper presented at the Goldschmidt Conference, Mass-Independent Isotope Variations, 2005.

Bindeman, Ilya N. "The Secrets of Supervolcanoes." *Scientific American* 294 (2006): 36–43.

Carey, S. and H. Sigurdsson. "The Intensity of Plinian Eruptions." *Bulletin of Volcanology* 51 (1989): 59–89.

Carter, N.L., C.B. Officer, C.A. Chesner, and W.I. Rose. "Dynamic Deformation of Volcanic Ejecta from the Toba Caldera: Possible Relevance to Cretaceous/Tertiary Boundary Phenomena." *Geology* 14 (1986): 380–383.

Chesner, C.A. "Petrogenesis of the Toba Tuffs, Sumatra, Indonesia." *Journal of Petrology* 39 (1998): 397–438.

Chesner, C.A., W.I. Rose, A. Deino, R. Drake, and J.A. Westgate. "Eruptive History of the Earth's Largest Quaternary Caldera (Toba, Indonesia) Clarified." Geology 19 (1991): 200–203.

Christiansen, R.L., G.R. Foulger, and J.R. Evans. "Upper-mantle Origin of the Yellowstone Hotspot." Bulletin of the Geological Society of America 114 (2002): 1245–1256.

Christy-Vitale, Joseph. *Watermark: The Disaster that Changed the World and Humanity 12,000 Years Ago.* New York: Paraview Pocket Books, 2004.

CRWMS M&O. "Probablistic Volcanic Hazard Analysis for Yucca Mountain, Nevada," U.S. Department of Energy, REV 0, 1996.

Dar, A., A. Laor, and N.J. Shaviv. "Life Extinctions by Cosmic Ray Jets." Physical Review Letters 80 (1998): 5813–5816.

Dayton, Tian. *Trauma and Addiction: Finding the Cycle of Pain Through Emotional Literacy.* Deerfield Beach, Fla.: HCI Communications, 2000.

Decker, Robert W., and Barbara B. Decker. *Mountains of Fire: The nature of volcanoes.* New York, N.Y.: Cambridge University Press, 1991.

————. *Volcanoes, Third Edition.* New York: W.H. Freeman, 1997.

Dietz, R.S. "Continent and Ocean Basin Evolution by Spreading of the Sea Floor." *Nature* 190 (1961): 854–857.

Doniger, Wendy. *The Rig Veda.* New York: Penguin Classics, 1981.

Eliot, Alexander. *The Universal Myths: Heroes, Gods, Tricksters and Others.* New York: Meridian, 1996.

Ellis, Richard. *No Turning Back: The Life and Death of Animal Species.* New York: HarperCollins, 2004.

Erdoes, Richard and Alfonso Ortiz. *American Indian Myths and Legends.* New York: Pantheon Books, 1984.

Ewert, J.W., M. Guffanti, and T.L. Murray. "An Assessment of Volcanic Threat and Monitoring Capabilities in the United States: Framework for a National Volcano Early Warning System." U. S. Geological Survey Open-File Report 2005–1164, April 2005.

FEMA. "Coping With Disaster." *www.fema.gov.* 2006.

FEMA. "What To Do Before, During and After a Volcano." *www.fema.gov.hazard/dproc.shtm.* 2006.

Fialko, Y. "Interseismic Strain Accumulation and the Earthquake Potential on the Southern San Andreas Fault System." *Nature* 441 (2006): 968–971.

Firestone, Richard, Alan West, and Simon Warwick-Smith. *The Cycle of Cosmic Catastrophes: Flood, Fire and Famine in the History of Civilization*. Rochester, Vt.: Bear and Company, 2006.

Franklin, Benjamin. "Meteorological Imaginations and Conjectures." Memoirs of the Literary and Philosophical Society of Manchester, Second Edition, London. *www.dartmouth.edu/~volcano/ Fr373p77.html*.

Geotimes News Notes, "Ozone Link to Permian Extinction," October 2005, 2 pages.

Gualda, G.A., and A.T. Anderson. (2006) "The Pre-eruptive Texture of the Bishop Magma." Trans. AGU, 87(52), Fall Meet. Suppl., Abstract V24C-04.

Hagen, Edward H. "The Evolutionary Psychology FAQ." *www.anth.ucsb.edu/projects/human/epfaq/design.html*. 1999–2002.

Hamilton, Edith. *Mythology*. Boston, Mass.: Little Brown, 1942.

Hansen, J., A. Lacis, R. Ruedy, and M. Sato. "Potential Climate Impact of the Mount Pinatubo Eruption." Geophysical Research Letters, 19 (1992): 215–218.

Harpending, Henry C., Mark A. Batzer, Michael Gurven, Lynn B. Jorde, Alan R. Rogers, and Stephen T. Sherry. "Genetic Traces of Ancient Demography." PNAS 95, Februray 1998.

Henderson, Bobby. *The Gospel of the Flying Spaghetti Monster*. New York, N.Y.: Villiard, 2006.

Hess, H.H. "History of Ocean Basins." In *Petrologic Studies: A Volume to Honor A. F. Buddington*, edited by A.E.J. Engel et al., Geological Society of America (1962): 599–620.

Highwood, E.J. and D.S. Stevenson. "Atmospheric Impact of the 1783-1784 Laki Eruption: Part II Climatic Effect of Sulfate Aerosol." *Atmospheric Chemistry and Physics* 3 (2003): 1177–1189.

Houghton, B.F., C.J.N. Wilson, M.O. McWilliams, M.A. Lanphere, D.S. Weaver, R.M. Briggs, and S.M. Pringle. "Chronology and Dynamics of a Large Silicic Magmatic System: Central Taupo Volcanic Zone, New Zealand." *Geology* 23 (1995): 13–16.

Husen, S., S. Wiemer, and R.B. Smith. "Remotely Triggered Seismicity in the Yellowstone National Park Region by the 2002 Mw 7.9 Denali Fault Earthquake, Alaska." *Bulletin of the Seismological Society of America* 94 (2004): S317–S331.

"Information for Survivors of Natural Disasters." National Center for PTSD, Nov. 26, 2000.

Ingman, Max. "Evolution: Investigating Human Evolution – Mitochondrial DNA Clarifies Human Evolution." In *ActionBioScience.org*, American Institute of Biological Sciences, 2000–2007.

Izett, G.A., J.D. Obradovich, and H.H. Mehnert. "The Bishop Ash Bed and Some Older Closely Related Ash Beds in California, Nevada, and Utah." *U.S. Geological Survey Open-File Report* (1982): 82–584.

Jacobs, James G. "Paleoanthropology in the 1990s." In *www.jqjacobs.net*.

Janis, Pam. "Do Cells Remember?" *USA Weekend* Special Reports, May 22-24, 1998.

Johanson, Donald. "Origins of Modern Humans: Multiregional or Out of Africa." In ActionBioScience.org, American Institute of Biological Sciences, 2000–2007.

Johnston, M.J.S., D.P. Hill, A.T. Linde, J. Langbein, and R. Bilham. "Transient Deformation During Triggered Seismicity from the 28 June 1992 Mw =7.3 Landers Earthquake at long Valley Volcanic Caldera, California." *Bulletin of the Seismological Society of America* 85 (1995): 787–795.

Johnston, M.J.S., S.G. Prejean, and D.P. Hill. "Triggered Deformation and Seismic Activity under Mammoth Mountain in Long Valley Caldera by the 3 November 2002 Mw 7.9 Denali Fault Earthquake." *Bulletin of the Seismological Society of America* 94 (2004): S360-S369.

Jones, Lucile M. and Paul A. Reasenberg. "Some Facts about Aftershocks to Large California Earthquakes in California." *U.S. Geological Survey Open-File Report* (1996): 96–266.

Jones, M.T., S.J. Sparks, and P.J. Valdes. (2006), "The Climatic Impact of Supervolcanic Ash Blankets." Trans. AGU, 87(52), Fall Meet. Suppl., Abstract V23G-07.

Kerr, R.A. "A Volcanic Crisis for Ancient Life." *Science* 270 (1995): 27–28.

Keys, David. Catastrophe: An Investigation into the Origins of Modern Civilization. New York, N.Y.: The Ballantine Publishing Group, First American Edition, 2000.

Kious, W. Jacquelyne, and Robert I. Tilling. "This Dynamic Earth: The Story of Plate Tectonics." U.S. Department of Interior Publication. Washington, D.C.: GPO, 1996.

Larson, Debra Levey. "Rural America More Prepared For Disaster—Also More Vulnerable." EurekAlert, January 12, 2007.

Linde, A.T. and I.S. Sacks. "Triggering of Volcanic Eruptions." *Nature* 395 (1998): 888–890.

Lipman, P.W. "Subsidence of Ash-flow Calderas: Relation to Caldera Size and Magma Chamber Geometry." *Bulletin of Volcanology* 59 (1997): 198–218.

Lipowicz, Alice. "The Next Disaster: Are We Ready? A Special Report on 10 High-risk Cities." *Reader's Digest*, July 2006.

Lowenstern, J.B. (2006), "Imagining the Unimaginable: How do we Forecast and Predict a Large Caldera Eruption?" Trans. AGU, 87(52), Fall Meet. Suppl., Abstract V24C-08.

Lowenstern, J.B., R.B. Smith, and D.P. Hill. "Monitoring Super-Volcanoes: Geophysical and Geochemical Signals at Yellowstone and other Large Caldera Systems." *Philosophical Transactions of The Royal Society A* 364 (2006): 2055–2072.

Lynch, D.K. "The San Andreas Fault." Online Fault Zone Map and Photos. *http://geology.com/articles/san-andreas-fault.shtml*.

MacRae, Penny. "'Doomsday Vault' to Resist Global Warming Effects." Agence France Presse, February 9, 2007.

Mason, Ben G., David M. Pyle, and Clive Oppenheimer. "The Size and Frequency of the Largest Explosive Eruptions on Earth." *Bulletin of Volcanology* 66 (2004): 735–748.

Masturyono, McCaffrey, R., D.A. Wark, S.W. Roecker, Ibrahim Fauzi. "Distribution of Magma Beneath the Toba Caldera Complex, North Sumatra, Indonesia, Constrained by Three-Dimensional P Wave Velocities, Seismicity, and Gravity Data." *Geochemistry Geophysics Geosystems* 2 (2001): 1525–2027.

McClaskey, Thomas R. "Decoding Traumatic Memory Patterns at the Cellular Level." The American Academy of Experts in Traumatic Stress, *www.aaets.org*. 1998.

McCoy, Floyd W., and Grant Heiken. "The Late-Bronze Age Explosive Eruption of Thera (Santorini), Greece: Regional and Local Effects." *Geological Society of America*, Special Paper 345 (2000): 43–70.

McGuire, Bill. *Apocalypse: A Natural History of Global Disasters*. London: Cassell & Co., The Orion Publishing Group, 1999.

Mercogliano, Chris Kim and Debus. "Expressing Life's Wisdom: Nurturing Heart-Brain Development Starting With Infants—A 1999 Interview with Joseph Chilton Pearce." *Journal of Family Life* 5 (1999).

Miller, C.D. "Potential Hazards from Future Volcanic Eruptions in California." U.S. Geological Survey Bulletin 1847, 17 pages, 1989.

Miller, C.D., D.R. Mullineaux, D.R. Crandell, and R.A. Bailey. "Potential Hazards from Future Volcanic Eruptions in the Long Valley-Mono Lake Area, East-Central California and Southwest Nevada—a Preliminary Assessment." U.S. Geological Survey Circular 877, 10 pages, 1982.

Morgan, J.P., T.J. Reston, and C.R. Ranero. "Contemporaneous Mass Extinctions, Continental Flood Basalts, and 'Impact Signals': Are Mantle Plume-Induced Lithospheric Gas Explosions the Causal Link?" *Earth and Planetary Science Letters* 217 (2004): 263–284.

Newhall, C.G. and S. Self. "The Volcanic Explosivity Index (VEI): an Estimate of Explosive Magnitude for Historical Volcanism." *Journal of Geophysical Research* 87 (1982): 1231–1238.

Newman, A.V., T.H. Dixon, G.I. Ofoegbu, and J.E. Dixon. "Geodetic and Seismic Constraints on Recent Activity at Long Valley Caldera, California: Evidence for Viscoelastic Rheology." *Journal of Volcanology and Geothermal Research* 105 (2001): 183–206.

Pearsall, Paul, Gary Schwartz, and Linda Russek. "Organ Transplants and Cellular Memories." *Nexus Magazine*, Issue May/June 2005.

Pert, Candace. *Molecules of Emotion*. London: Simon and Schuster UK Ltd., 1999.

Rampino, M.R. and S. Self. "Historic Eruptions of Tambora (1815), Krakatau (1883), and Agung (1963), Their Stratospheric Aerosols and Climatic Impact." *Quaternary Research* 18 (1982): 127–143.

———. "Volcanic Winter and Accelerated Glaciation Following the Toba Super-eruption." *Nature* 359 (1992): 50–52.

———. "Climate-Volcanism Feedback and the Toba Eruption of ~74,000 Years Ago." *Quaternary Research* 40 (1993): 269–280.

Rampino, M.R. and R.B. Stothers. "Terrestrial Mass Extinctions and Galactic Plane Crossings." *Nature* 313 (2002): 159–160.

Rampino, M.R. and Stephen Self. "Climate-Volcanism Feedback and the Toba Eruption of ~74,000 Years Ago." *Quaternary Research* 40 (1993): 269–280.

Rampino, Michael R. and Stanley H. Ambrose. "Volcanic Winter in the Garden of Eden: The Toba Supereruption and the Late Pleistocene Human Population Crash." Geological Society of America, Special Paper 345, (2000): 71–82.

Rampino, Michael. "Supereruptions as a Threat to Civilisations on Earth-like Planets." *Icarus* 156 (2002): 562–56.

Reid, G.C., J.R. McAfee, and P.J. Crutzen. "Effects of Intense Stratospheric Ionization Events." *Nature* 275 (1978): 489–492.

Robock, A. "Volcanic Eruptions and Climate." *Reviews of Geophysics* 38 (2000): 191–207.

Robock, A. and C. Mass. "The Mount St. Helens Volcanic Eruption of 18 May 1980: Large Short-Term Surface Temperature Effects." *Science* 216 (1982): 628–630.

Robock, A., C. Ardmann, L. Oman, D. Shindell, and G. Stenchikov. (2006), "Can Volcanic Eruptions Produce Ice Ages or Mass Extinctions?" Eos Trans. AGU, 87(52), Fall Meet. Suppl., Abstract V23G-06.

Rogers, Alan R. and Lynn B. Jorde. "Genetic Evidence on Modern Human Origins." *Human Biology 67* (1995).

Rose, W.I. and C.A. Chesner. "Worldwide Dispersal of Ash and Gases from Earth's Largest Known Eruption: Toba Sumatra, 75 ka." *Global and Planetary Change* 89 (1990): 269–275.

Russell, D.A. and W. Tucker. "Supernovae and the Extinction of the Dinosaurs." *Nature* 229 (1971): 553–554.

Ryskin, G. "Methane-Driven Oceanic Eruptions and Mass Extinctions." Geology 31: (2003): 741–744.

Self, S., R. Gertisser, T. Thordarson, M.R. Rampino, and J.A. Wolff. "Magma Volume, Volatile Emissions, and Stratospheric Aerosols from the 1815 Eruption of Tambora." *Geophysical Research Letters*, 31 (2004): L20608. (doi: 1029/2004GL020925).

Self, S., J.X. Zhao, R.E. Holasek, R.C. Torres, and A.J. King. "The Atmospheric Impact of the Mount Pinatubo Eruption, Fire and Mud: Eruptions and Lahars of Mount Pinatubo, Philippines." (eds. Newhall, C. G. and Punongbayan, R. S.), Philippine Institute of Volcanology and Seismology; University of Washington Press, (1996): 1089–1115.

Self, Stephen. (2006) "Defining Super-eruptions: Purpose, Prejudices, and Limitations." Eos Trans. AGU, 87(52), Fall Meet. Suppl., Abstract V23G-02.

———. "The Effects and Consequences of Very Large Explosive Volcanic Eruptions." *Philosophical Transactions of the Royal Society A* 364 (2006): 2073–2097.

Sigurdsson, Haraldur, Bruce Houghton, Stephen R. McNutt, John Stix, and Haxel Rymer. *Encyclopedia of Volcanoes*. New York: Academic Press, 2000.

Simkin, Tom and Lee Siebert. *Volcanoes of the World: A Regional Directory, Gazetteer, and Chronology of Volcanism during the Last 10,000 Years*. Tuscon, Az.: Geoscience Press, 1994.

Solomon, Susan. "Stratospheric Ozone Depletion: a Review of Concepts and History." *Reviews of Geophysics* 37 (1999): 275–316.

Stothers, R.B. "The Great Tambora Eruption of 1815 and its Aftermath." *Science* 224 (1984): 1191–1198.

Stothers, R.B. "The Mystery Cloud of A.D. 536." *Nature* 307 (1984): 344–345.

Takeuchi, Leslie A. "Cellular Memory in Organ Transplants." *San Francisco Magazine*, August 2000.

Than, Ken. "NASA Tests Inflatable Lunar Shelters." Space.com, March 28, 2007.

U.S. Geological Survey. "Carbon Dioxide and Helium Discharge from Mammoth Mountain." Long Valley Observatory, 2001.

———. "Invisible CO2 Gas Killing Trees at Mammoth Mountain, California." Fact Sheet 172–96, Online Version 2.0, 1996.

———. "Living with a Restless Caldera-Long Valley, California." Fact Sheet 108–96, Online Version 2.1, revised May 2000.

———. "Steam Explosions, Earthquakes, and Volcanic Eruptions-What's in Yellowstone's Future?" Fact Sheet 2005–3024, 2005.

———. Future Eruptions in California's Long Valley Area—What's Likely." Fact Sheet 073–97, Version 1.1, 1997.

———. "What to Do If A Volcano Erupts" informational series. *http://volcan.wr.usgs.gov/Hazards/ Safetywhat_to_do_during_ashfall.html.* 2006.

Voorhies, Mike. "Ashfall: Life and Death at a Nebraska Waterhole Ten Million Years Ago." University of Nebraska State Museum, Museum Notes Number 81, February 1992.

Waite, G.P., R.B. Smith, and R.M. Allen. "V_p and V_s Structure of the Yellowstone Hot Spot from Teleseismic Tomography: Evidence for an Upper Mantle Plume." *Journal of Geophysical Research* 111, B04303, doi: 10.1029/2005JB003867

Ward, Peter. "Impact from the Deep." *Scientific American* 295 (2006): 64–71, 2006.

Weber, George, "Toba Volcano." Published for the Andaman Association, *www.andaman.org.* 2006.

Weldon, R.J., T.E. Fumal, G.P. Biasi, and K.M. Scharer. "Past and Future Earthquakes on the San Andreas Fault." *Science* 308 (2005): 966–967.

Whitehouse, David. "Genetic Study Roots Humans in Africa." BBC News, Dec. 6, 2000.

———. "Humans Came 'Close to Extinction.'" BBC News, Sept. 8, 1998.

———. "When Humans Faced Extinction." BBC News, Sept. 19, 2006.

Wignall, P.B. "Large Igneous Provinces and Mass Extinctions." *Earth-Science Reviews* 53 (2001): 1–33.

Wignall, P.B. and R.J. Twitchett. "Oceanic Anoxia and the End Permian Mass Extinction." *Science* 272 (1996): 1155–1158.

Wilson, C.J. (2006), "Insights from Field Geology into the Styles and Timings of Large Silicic Explosive 'Supereruptions.'" Eos Trans, AGU, 87(52), Fall Meet. Suppl., Abstract V23G-01.

Wilson, J. Tuzo. "A New Class of Faults and Their Bearing on Continental Drift." *Nature* 207 (1965): 343–347.

Winchester, Simon. *Krakatoa: The Day the World Exploded: August 27, 1883.* London: Penguin Viking, New York, N.Y.: HarperCollins, 2003.

Woods, A.W. and Ken H. Wohletz. "Dimensions and Dynamics of Co-ignimbrite Eruption Columns." *Nature* 350 (1991): 225–227.

Zeilinga de Boer, Jelle and Donald T. Sanders. "Volcanoes in Human History: The Far Reaching Effects of Major Eruptions." Princeton, New Jersey and Oxford. Princeton University Press, 2002.

Resources

USGS
U.S. Geological Survey
www.usgs.gov
1-800-ASK-USGS

VOLCANO WATCH
http://hvo.wr.usgs.gov/volcanowatch/

GLOBAL VOLCANISM PROGRAM
Smithsonian National Museum of Natural History
www.volcano.si.edu

THE ANDAMAN ASSOCIATION
www.andaman.org
George H.J. Weber-President

The Andaman Association
Waldstrasse 6
CH-4410 Liestal Switzerland

VOLCANO WORLD
http://volcano.und.edu/

VOLCANOES.COM
www.volcanoes.com/

ASHFALL FOSSIL BEDS STATE HISTORICAL PARK
http://ashfall.unl.edu/
Rick Otto, Superintendent
86930 517th Avenue
Royal, NE 68773
Phone: (402) 893-2000
E-mail: ashfall2@unl.edu

GODDARD INSTITUTE FOR SPACE STUDIES
www.giss.nasa.gov/

BENFIELD GREIG HAZARD RESEARCH CENTRE
www.benfieldhrc.org

WEBSITES

MICHAEL RAMPINO
www.nyu.edu/fas/biology/faculty/rampino/index.html

STEPHEN SELF
www3.open.ac.uk/Earth-Sciences/people/53.shtml

STANLEY AMBROSE

www.anthro.uiuc.edu/faculty/ambrose/
www.bradshawfoundation.com/stanley_ambrose.php

Index

W

water supply, ash in the, 217
Weak Garden of Eden, *see Weak GEO*
winter, volcanic, 227
world, volcanoes around the, 82-91

Y

Yellowstone
    eruptions, 194-199
    Hot Spot, 193
    National Park, map of, 198
    Volcano Observatory, *see YVO*
YVO, 243

About the Authors

DR. JOHN SAVINO, PH.D.

Dr. John Savino graduated magna cum laude from Fairfield University in 1962.

He received a master of arts degree in physics and a doctor of philosophy degree in geophysics (major in seismology) from Columbia University, New York. Graduate studies for the Ph.D. and post-graduate studies were performed while at the Lamont-Doherty Earth Observatory of Columbia University. While at Columbia University, research areas included analyses of the spatial and temporal occurrence of great earthquakes in light of plate tectonics, development of techniques for discrimination between worldwide earthquakes and underground nuclear explosions, development and application of laser technology to broadband seismic recording, and development of high-gain instrumentation for detection of very small earthquakes and underground explosions. Since 1991, Dr. Savino has been involved in the High-Level Nuclear Waste disposal project proposed for Yucca Mountain, Nevada. He has assisted Department of Energy (DOE) project managers in reviewing and assessing research conducted by earth scientists at several national laboratories and universities relating to the seismic and volcanic hazards in the Yucca Mountain region. Dr. Savino has participated in the DOE's Public Outreach Program delivering presentations on earthquakes and volcanoes and has presented papers at national and international scientific conferences.- He has also published numerous articles in refereed journals, technical reports and abstracts in conference and meeting programs. He is an active member of the American Geophysical Union. Dr. Savino divides his time between residences in Big Bear Lake, California, and Las Vegas, Nevada.

MARIE D. JONES

Marie D. Jones is the widely published author of *PSIence: How New Discoveries in Quantum Physics and New Science May Explain the Existence of Paranormal Phenomena* (New Page Books, November 2006) and *Looking For God In All the Wrong Places* (Paraview, 2003), chosen as the "Best Spiritual/Religious Book of 2003" by the popular book review Website, *RebeccasReads.com*, and the book made the "Top Ten of 2003" list at *MyShelf.com*. She has coauthored more than 30 inspirational books for PIL/New Seasons, including *Life Changing Prayers*, *100 Most Fascinating People in the Bible*, *Right Words for Right Times* and many others. An accomplished screenwriter, she has written and produced videos for national distribution, including an award-winning children's storybook video, and is credited with more than 300 published magazine and Internet articles, essays, and stories. Her work has been featured in prominent anthologies, including several *Chicken Soup for the Soul* books, *Rocking Chair Reader*, *If Women Ruled the World*, *God Allows U-Turns*, and *Angels on Earth*, and she is a popular book reviewer for several online review Websites, including *CurledUp.com* and *BookIdeas.com*. She is also a licensed New Thought Metaphysics, minister and pastoral counselor with a background in religious studies, metaphysics and the paranormal, including 15 years as a trained field investigator for the Mutual UFO Network, and she has written for MUFON Journal, *UFO Magazine*, *InnerSelf*, Beyond Reality, and other paranormal and new science publications. She is an avid student of quantum physics, Earth sciences, new science, and consciousness research. She is a Toastmasters-trained public speaker and winner of many regional Best Speaker awards. Marie D. Jones lives in San Marcos, California.